Environmental, Ethical, and Economical Issues of Nanotechnology

Environmental, Ethical, and Economical Issues of Nanotechnology

edited by

Chaudhery Mustansar Hussain
Gustavo Marques da Costa

JENNY STANFORD
PUBLISHING

Published by

Jenny Stanford Publishing Pte. Ltd.
101 Thomson Road
#06-01, United Square
Singapore 307591

Email: editorial@jennystanford.com
Web: www.jennystanford.com

British Library Cataloguing-in-Publication Data
A catalogue record for this book is available from the British Library.

Environmental, Ethical, and Economical Issues of Nanotechnology

ISBN 978-981-4877-76-3 (Hardcover)
ISBN 978-1-003-26185-8 (eBook)

Contents

Preface   xi

1. **Nanomaterials in the Environment: Definitions, Characterizations, Effects, and Applications**   **1**

   Daniela Patrícia Freire Bonfim,
   Gabriela Brunosi Medeiros,
   Alessandro Estarque de Oliveira,
   Vádila Giovana Guerra, and Mônica Lopes Aguiar

   1.1   Nanotechnology and Nanomaterials:
   Definitions and Properties   2
   1.2   Nanotechnology and Nanomaterials:
   Applications   7
   1.3   Characterization Techniques of Nanomaterials   9
         1.3.1   Scanning Electron Microscopy   9
         1.3.2   Atomic Force Microscopy   10
         1.3.3   Spectroscopic Analysis   10
         1.3.4   Dynamic Light Scattering   11
         1.3.5   X-ray Diffraction   11
   1.4   Risks to Human Population and Environment   12
   1.5   Nanoscale Air Pollutants   13
   1.6   Fibrous Filters and Nanofiltration   14
         1.6.1   Theory of Air Filtration   15
         1.6.2   Air-Filtration Applications:
   Nanofibers   20
   1.7   Final Considerations and Perspectives   22

2. **Nanofiber Production Techniques Applied to Filtration Processes**   **31**

   Alessandro Estarque de Oliveira, Daniela Patrícia Freire
   Bonfim, Ana Isabela Pianowski Salussoglia,
   Gabriela Brunosi Medeiros, Vádila Giovana Guerra,
   and Mônica Lopes Aguiar

   2.1   Production Techniques of Nanofibers and
   Nanostructures   32

	2.1.1	Drawing	33
	2.1.2	Electrospinning	34
	2.1.3	Co-electrospinning and Solution Blow Spinning	36
	2.1.4	Centrifugal Spinning	37
	2.1.5	Melt Spinning and Melt Blow Spinning	38
	2.1.6	Dry Spinning, Wet Spinning, and Gel Spinning	40
	2.1.7	Template Synthesis	41
	2.1.8	Phase Separation	44
	2.1.9	Self-Assembly	45
2.2		Techniques of Characterization	46
2.3		Conclusion	49

3. Risks and Effects on Human Health of Nanomaterials 61

Gustavo Marques da Costa, Aline Belem Machado,
Daniela Montanari Migliavacca Osório,
Daiane Bolzan Berlese, and
Chaudhery Mustansar Hussain

3.1		Introduction	62
	3.1.1	Nanomaterials and Their Effects	62
3.2		Toxicity, Transport, and Destination	62
3.3		Risk of Nanoparticles to Humans	65
3.4		Conclusion	69

4. Bioavailability and Toxicity of Manufactured Nanoparticles in Terrestrial Environments 73

Aline Belém Machado, Gustavo Marques da Costa,
Daniela Montanari Migliavacca Osório,
Daiane Bolzan Berlese, and
Chaudhery Mustansar Hussain

4.1		Origin of Manufactured Nanoparticles in Terrestrial Environments	74
4.2		Bioavailability of Nanomaterials in Terrestrial Environments	77
	4.2.1	Aggregation/Agglomeration	80
	4.2.2	Dissolution and Redox	80
	4.2.3	Adsorption	82
	4.2.4	Chemical Processes	82

4.3	Toxicity of Manufactured Nanoparticles	82
4.4	Conclusion	83

5. Occupational Health Hazards of Nanoparticles   **89**

Sandra Magali Heberle and Michele dos Santos
Gomes da Rosab

5.1	Introduction	89
5.2	Anatomy of the Respiratory System	95
	5.2.1 Respiratory Control Breathing	96
	5.2.2 Lung Capacity and Lung Volumes	96
	5.2.3 Spirometry	97
	5.2.4 Basic Standards to be Recognized in Spirometry	100
5.3	Nanomedicine and Pulmonary Disease Therapy	104
5.4	Nanoparticles Benefits in Nasal and Inhalation Therapy	104

6. Ethical Issues in Nanotechnology-I   **109**

Maurício Machado da Rosa

6.1	Measures and Indicators of Societal Impacts	110
6.2	Societal Implications of Nanoscience and Nanotechnology	113
6.3	Social Acceptance of Nanotechnology	114
6.4	Identifying Ethical Issues	117
	6.4.1 General Ethical Concepts	117
	6.4.2 Ethical Principles	118
	6.4.3 Ethical Issues of Modified Foods	119
	6.4.4 Medical Ethical Issues	119
	6.4.5 Usability Ethical Issue	120
	6.4.6 Implanted Nanochip Ethical Issues	121
6.5	Addressing the Issues to the Geographical, Economic, Psycho-Social, Affective, Cognitive, Technical Administrative, and Political Parameters	121
	6.5.1 Geographical Issues	122
	6.5.2 Economic Issues	123
	6.5.3 Psycho-Social, Affective, and Cognitive Issues	123

6.5.4	Technical Administrative Issues	124
6.5.5	Governance and Political Parameters	125
6.6	Ethical, Religious, and Cultural Acceptability	126

7. Ethical Issues in Nanotechnology-II — **131**

*Wilson Engelmann, Raquel von Hohendorff,
and Daniele Weber da Silva Leal*

7.1	Introduction	132
7.2	Insertion of Nanotechnologies into Human Life and the Risks: The Emergence of an Ethico-Legal Category	135
7.3	Shifting from "The Coherent Life Plan" (Finnis) Model to the "Coherent Ethico-Legal Plan" Model to Regulate Challenges Brought about by Nanomaterials	140
7.4	Final Considerations	145

8. Economics of Nanotechnology-I and Modern Policy and Decision-Making about Nano — **151**

*Gustavo Marques da Costa and
Chaudhery Mustansar Hussain*

8.1	Introduction	151
8.2	Public and Private Investments in Nanotechnology	152
8.3	Nanotechnology and Unintended Consequences	154
8.4	Nano-Mechanisms as Unique Governance Challenges: Nanotechnology and Security	155
8.5	Planning for the Unexpected	157
8.6	Vision of Modern Life with Nano	157
8.7	Nanotechnology and Its Interfaces	160
8.8	Final Considerations	161

9. Economics of Nanotechnology-II — **165**

*Michele dos Santos Gomes da Rosa and
Maurício Machado da Rosa*

9.1	Nanotechnology (Benefits and Risks)	166
9.2	Community Ownership	166
9.3	Nano-Infrastructure	167

9.4 Commercialization of Nanotechnologies   167
9.5 National Security/Economic Competitiveness   168
9.6 Nanomedicine and Human Body   169
9.7 Economic Impact   170
9.8 Sustainability: Environment   171

10. Legalization of Nanotechnology   **177**

Maicon Artmann, Roberta Verdi, Vanusca Dalosto
Jahno, and Haide Maria Hupffer

10.1 Introduction   178
10.2 Regulatory and Governance Initiatives for
      Nanotechnology in the European Union   178
10.3 An Overview of Regulatory Initiatives in
      the United States and China   186
10.4 Contribution of Intergovernmental
      Organizations to the Governance or Risk
      Management of Nanotechnologies   191
10.5 Conclusion   198

11. Future: Green and Sustainable Nano   **205**

Gustavo Marques da Costa and
Michele dos Santos Gomes da Rosa

11.1 Future   206
      11.1.1 The Application   209
11.2 Conclusion   215

Index   219

Preface

Nano is moving out of its comfort zone of scientific discourse. As new products go to market and national and international organizations roll out public engagement programs on nanotechnology to discuss environmental and health issues, various sectors of the public are beginning to discuss what all the controversy is about. Nongovernmental organizations have long since reacted; however, now the social sciences have begun to study the cultural phenomenon of nanotechnology, thus extending discourses and opening out nanotechnology to whole new social dimensions. Social dimensions and their new constructed imaginings around nanotechnology intersect with the economy, ecology, health, governance, and illusory futures. There is always a need for more than just an ELSI sideshow within nanotechnology. The collective public imaginings of nanotechnology include tangles of science and science fiction, local enterprise, and global transformation, all looking forward toward a sustainable future while looking back on past debates about science and nature. Nanotechnology is already very much embedded in the social fabric of our life and times. This book addresses these new challenges of nanotechnology in detail with an up-to-date knowledge on environmental health and economical concerns of nano.

Economic value of nanotechnology is measured in terms of employment, education, research activity, and commercialization of products and processes. These metrics can vary widely, therefore, it is difficult to define the base metrics of nanotechnology. Patents and published articles and papers provide a useful metric, along with research projects and outputs, to gauge the value of academic activity. However, economic metrics are more complex. Evaluating the return on investment from nanotechnology investments is much more difficult for government agencies and policymakers. A government can justify the value derived from millions or billions of dollars in nanotechnology investments by evaluation in terms of job creation, reduction of manufacturing costs, new company formation, contribution to export industries, and creation of new products or services—but there is no straightline metric that can interpret public funding of nanotechnology initiatives directly into commercial value. Therefore, a detailed assessment of the business potential of nanotechnology applications, based on nanomaterials, nanotools, and nanodevices is always required. Moreover, the benefits of incorporating nanomaterials in commercial products and processes will bring challenges with them for environmental, health, and safety risks; ethical and social issues; as well as uncertainty concerning market and consumer acceptance. Therefore, there is a need to establish relationships in nanoeconomy and opportunities for nanoapplications as well as to critically analyze their practical challenges, especially related to their impacts on the health and safety of workers involved in this innovative sector. The target of this book is to deliver a comprehensive and an easy-to-read text for anyone working in the field of environmental health and economical concerns of the nanoarena and to provide essential information to consultants and regulators about nanotechnology applications and processes helpful in their evaluation and decision-making procedures.

College and university graduates and postgraduate students taking advanced level courses on nanoscience and nanotechnology will find this book highly up-to-date, easy to use, and understandable. This book will ease their thirst of learning of environmental, health, and economical concerns of nanomethodologies and nanotechniques. Therefore, this book is mainly aimed for advanced degree–level students in nanoscience and nanoengineering. Major safety risks associated with nano are basically uninformed nanomaterial hazardous properties as well as the difficulties in characterizing exposure and defining emerging risks for the workforce. Various action strategies are required for the assessment, management, and communication of risks, aimed at the precautionary adoption of preventive measures, including hiring and training of employees, development of collective and personal protective equipment, and implementation of health surveillance programs to protect the health and safety of nano workers, which are necessary for occupational health considerations to have sustainable development of nanotechnology. We hope this book will prove to be a major milestone in the field of environmental health and economical concerns of nano.

The book is intended for a broad audience, working in the fields of nano (science, technology, and engineering) materials science, green chemistry, sustainability, device engineering, etc. It will be an invaluable reference source for libraries of universities and industrial institutions, government and independent institutes, and individual research groups and scientists working in the field of nano. Have a great reading!

Chaudhery Mustansar Hussain
Gustavo Marques da Costa
February 2022

Chapter 1

Nanomaterials in the Environment: Definitions, Characterizations, Effects, and Applications

Daniela Patrícia Freire Bonfim, Gabriela Brunosi Medeiros, Alessandro Estarque de Oliveira, Vádila Giovana Guerra, and Mônica Lopes Aguiar

Graduate Program of Chemical Engineering, Federal University of São Carlos, Rodovia Washington Luís, km 235, PO Box 676, Zip Code 13560-970, São Carlos, SP, Brazil

mlaguiar@ufscar.br

Nanoscience and technology have grown rapidly in recent years and have already had a major influence on the development of new materials worldwide, consolidating nanotechnology as a science that promises to be the main economic engine in the near future. For a better understanding of this new technology, the objective of this chapter is to address the current scenario of nanomaterials from the presentation of important concepts related to their definition, application, and characterization, with emphasis on the development and use of nanofibers in the remediation of air pollution.

Environmental, Ethical, and Economical Issues of Nanotechnology

Edited by Chaudhery Mustansar Hussain and Gustavo Marques da Costa

Copyright © 2022 Jenny Stanford Publishing Pte. Ltd.

ISBN 978-981-4877-76-3 (Hardcover), 978-1-003-26185-8 (eBook)

www.jennystanford.com

1.1 Nanotechnology and Nanomaterials: Definitions and Properties

Nanotechnology refers to the science and technology of the design, construction, manipulation, and understanding of nanoscale materials and systems, which has currently demonstrated significant impacts on several applications in the biomedical field, environment, and energy [15, 36, 60, 63]. Therefore, the term nanotechnology is described by the technological applications of devices or materials that have one or more of their physical dimensions less than 100 nm (billionth of a meter) [11, 17]. The first milestone of nanotechnology was in 1959, when the American physicist Richard Feynman, in one of his lectures entitled "There's Plenty of Room at the Bottom," showed that in the atomic dimension, works with different laws and, thus, new types of effects and possibilities, suggesting material manipulation on a nanometric scale [24]. In 1974, Norio Taniguchi coined the term "nanotechnology" to label precision machining with a tolerance of a micron or less [78].

In 1981, Gerd Binnig and Heinrich Rohrer developed Scanning Tunneling Microscope (STM), which allowed atomic mapping of a material. The principle of operation of this microscope is a metal probe that scans the surface of a sample to determine surface topography [4]. It was from then on that other types of electron microscopes, such as the atomic force microscope (AFM), derived from STM and which allowed mapping non-conductive materials, appeared.

In 2005, Technical Committee of the International Organization for Standardization (ISO/TC 229) was created with the purpose of standardizing concepts in the field of nanotechnology in relation to the understanding and control of materials and processes on a nanoscale and the use of their exclusive properties for the creation of new materials, equipment, and systems. The specific objectives of this committee include the determination of nomenclatures and terminologies, aspects related to metrology and instrumentation, testing methodologies, modeling, simulations, and practices associated with health, safety, and the environment [11, 29, 32].

The definitions related to nanotechnology were presented in a Technical Specifications of the International Organization for Standardization (ISO/TS 80004). Among the concepts covered in this committee is the definition for nanomaterials: "Material with any external dimension in the nanoscale or having internal structure or surface structure in the nanoscale. This generic term is inclusive of nano-object and nanostructured material." The ISO/TS 27687 prepared by the European Committee for Standardization (CEN) defines the terminology for some types of nano-objects, including six distinct forms and an additional specific case (the quantum dot), shown in Table 1.1 [11, 29–32].

Table 1.1 Definition of terminology for some types of nano-objects, with their respective illustrated shapes

Nanoparticle	Nano-object with three external dimensions in the nanoscale.	
Nanoplate	Nano-object with three external dimensions in the nanoscale and the two other external dimensions significantly larger.	
Nanofiber	Nano-object with two similar external dimensions in the nanoscale and the third dimension significantly larger.	
Nanotube	Hollow nanofiber.	
Nanorod	Solid nanofiber.	
Nanowire	Electrically conducting or semiconducting nanofiber.	
Quantum dot	Crystalline nanoparticle that exhibits size-dependent properties due to quantum confinement effects on the electronic states.	

Notably, nanotechnology is on the rise, as it makes it possible to explore the physical–chemical and biological properties of structures in sizes between atoms and molecules [37, 50].

The origin of nanomaterials can be natural and anthropogenic or incidental (generated as a by-product of anthropogenic processes). Nanomaterials of natural origin are of biological origin, such as viruses and bacteria, and mineral origin, coming from mineral dust, volcanic rocks, etc. Anthropogenic or incidental ones are those produced from human activities, unintentional and with no specific purpose, as in the processes of refining, welding, food production, or automobile combustion. There are also manufactured nanomaterials, which are developed by people for a specific purpose. Thus, knowledge of the sources of nanoparticles is essential to understand their physical–chemical properties and to predict possible effects on the environment and human health [7, 9, 44].

The development of nanomaterials can be approached in two main ways: bottom-up and top-down. In the bottom-up approach, the materials are developed from molecular components that are chemically organized, forming a material on a nanoscale. In the top-down approach, nanomaterials are built from others of macroscopic scale. The main problem with the top-down approach is the imperfection of the surface structure, so the nanomaterials produced by the bottom-up approach are in a state closer to a thermodynamic equilibrium [5].

As can be seen in Fig. 1.1, the synthesis, structure and composition, properties, and application of a material are interconnected. Therefore, it is essential to know the structure and properties of the individual particles that make up this material to develop an adequate synthesis technique for the production of this material, seeking the desired properties for a given application and analyzing the possible effects of these materials on the environment [83].

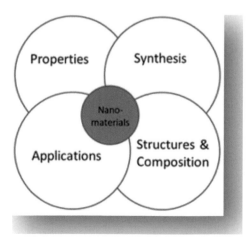

Figure 1.1   Synthesis, structure and composition, properties, and application of nanomaterial interconnections.

The structure of a nanomaterial can be divided into two components: the crystalline part with nanometric dimensions, which relates to the massive material (bulk); and the interphase, composed of defects, such as gaps, disagreements, grain contour, among others [20]. Solids with nanometric particle sizes are considered intermediate species between individual molecules and bulk, and so they present a distinction in their properties [66]. Depending on the environmental conditions (temperature and pressure), in addition to the technique used to produce a nanomaterial, it is possible to structurally modify a nanomaterial. Therefore, the crystalline structural organizations of these nanomaterials may differ in important physical and chemical properties, which specifies the application of this material [70, 83].

Besides, the different crystalline structures of the nanomaterial can affect its behavior and toxicity and should also be considered when analyzing possible adverse effects on organisms or environmental behavior [12, 57, 58].

The physical properties of a solid (diffusion, solubility, specific heat, entropy, thermal expansion, optical and infrared absorption, magnetic, electrical, mechanical) depend mainly on the size and chemical composition of the atomic structure. Thus, nanostructured materials may have new physicochemical properties, related to the effects of size reduction [62].

Upon reaching nanometer scales, a material shows changes in its properties due to mainly two factors: surface effects and quantum confinement. The surface effects are due to the high surface area/volume ratio, since nanomaterials have more surface atoms, which can participate in all physical and chemical interactions of the material with the medium in which it is inserted, when compared to macroscopic materials [9]. Quantum confinement is due to the behavior of atoms and electrons in matter. The wave properties of electrons in a material are influenced by variations in the scale, so a material developed at the nanoscale may show variations in physical, optical, magnetic, and mechanical properties, without changing its chemical composition [18, 77].

Concerning surface effects, the increase in the surface area of nanoscale materials makes them attractive for application in the environmental area as catalysts in photocatalytic activity and in the adsorption of polluting gases, in comparison with conventional materials [35, 59]. However, the physical–chemical and biological characteristics of these materials (such as size, shape, chemistry, surface properties, agglomeration, solubility, charge and effects of linked functional groups, and crystalline structure) make them potentially more reactive than their macro-sized equivalents [9, 17, 18]. Thus, in recent years, it has become necessary to study the toxicity of nanomaterials, both for the environment and human health [12, 57, 58, 70].

Nanostructured materials are classified as zero-dimensional (0D), one-dimensional (1D), two-dimensional (2D), and three-dimensional (3D) nanostructures [32, 62]. As seen in Fig. 1.2, 0D nanomaterials are composed of materials that have all their dimensions at the nanoscale, such as nanoparticles, quantum dots, and fullerenes; 1D are composed of materials that present two of their nanoscale dimensions, such as nanofibers, nanowires, and nanorods; 2D are composed of materials that have only one of their dimensions at the nanoscale, such as nanofilms, nanoplates, and networks; and 3D are materials that are not confined to the nanoscale in any dimension, but they may contain nano-sized grains, such as porous mesh [62].

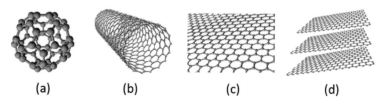

(a)   (b)   (c)   (d)

Figure 1.2 Characteristic dimensions of nanomaterials: (a) 0D, (b) 1D, (c) 2D, and (d) 3D.

Shehzad et al. [69] showed a study of syntheses that transform 2D materials into 3D. Three-dimensional materials are very complex and sophisticated, which have unique properties that make them applicable in several areas, such as energy, environment, and bio-related fields. The author also mentions that nanomaterials can be synthesized from inorganic, organic, carbon, and polymeric materials.

1.2 Nanotechnology and Nanomaterials: Applications

Therefore, as seen previously, nanomaterials can also be classified according to the material to which they were synthesized. Given this, one can mention carbon-based nanomaterials, nanocomposites, metals and alloys, biological nanomaterials, nanopolymers, and nanoceramics [32].

Carbon-based materials are common and versatile, as examples we can mention diamond, graphite, and fullerene, which exhibit different properties and structures. Carbon nanotubes, carbon nanofibers, black carbon, graphene flakes, and fullerenes are nanomaterials that have been used in various applications, such as biomedicine, energy, and sensors, in addition to contributing to environmental issues, such as pollution control, absorbing metals, antibiotics, and harmful gases. However, these nanomaterials are harmful to living organisms in the environment and are difficult to degrade under natural conditions, which has prompted several studies to assess the toxicity and degradation of carbon-based nanomaterials [35, 57].

Most nanopolymers are at an early stage of development, and the applications described are still in the research and development phase. Among the materials developed are nanofibers, hollow nanofibers, core–shell nanofibers, and nanobands or nanotubes that have great potential for applications, including homogeneous and heterogeneous catalysis, sensors, filter applications, and optoelectronics [21].

Due to the diversity of materials that can be developed in a nanoscale, as well as the peculiar properties of each nanomaterial, there is a wide field of application, such as medicine, electronics, computer science, physics, chemistry, biology, and engineering [7, 69].

In the field of medicine, there is the development of new drug diffusion systems that reach specific points in the human body (drug delivery); prostheses biocompatible with human organs and fluids; materials for tissue engineering and scaffolds, all from nanomaterials [65, 68, 71].

Addressing electronics and computational science, data recording using means that use nanolayers and quantum dots; wireless technologies; new devices and processes within all aspects of information and communication technologies; increased data processing speeds and storage capacities are some examples of the use of nanotechnology in search of new materials [69, 75].

The search for nanocatalysts has increased the energy efficiency of chemical transformation plants and also increased the combustion efficiency of motor vehicles (thus reducing pollution) in the chemical and materials industry [7].

New types of batteries; safe storage of hydrogen for use as a clean fuel; energy savings resulting from the use of lighter materials; and increasingly smaller circuits are among the studies developed with nanomaterials in the energy sector [7, 80].

In the environment, studies with selective membranes that can filter contaminants; nanostructured devices, capable of removing pollutants from industrial effluents; characterization of the effects of nanostructures on the environment; significant reduction in the use of materials and energy; reduction in sources of pollution; and triboelectric air filtering nanotechnology are being developed with nanomaterials [7, 12, 42, 57, 70].

The use of nanomaterials to control air pollution has gained focus, since the laws are increasingly stricter on the emission of polluting gases and particulate matter (PM). One of the recent examples is polymeric nanofibers used as air filters. Studies show that they can remove PM of 0.3 μm, with 99.998% efficiency, making them potentially applicable for air filtration. Moreover, the nanofiber filters showed properties of superhydrophobicity, desirable transparency (91%), and long-term stability [41]. Another example is the use of nanomaterials as catalysts for air pollution, such as for the abatement of NO_X gases (nitrogen monoxide (NO) and nitrogen dioxide (NO_2)) and volatile organic compounds, and also the oxidation of carbon monoxide [45, 59].

Recent studies show the use of nanofibers to apply filters with antibacterial action [12, 57]. Liu et al. [42] addressed a new nanotechnology, called triboelectric filtering. The authors showed that this type of filter can be applied in air purification in stores, hotels, supermarkets, shopping centers, and hospitals, since it can efficiently remove particulate material and kill bacteria, viruses, and other microorganisms. Selvam and Nallathambi [67] use polymeric nanofibers (polyacrylonitrile (PAN)) incorporated with silver nanoparticles (Ag) to make a protective mask with antibacterial activity.

1.3   Characterization Techniques of Nanomaterials

Given the potential application of nanomaterials, it is essential to know their characteristics and properties, as previously mentioned. For this purpose, the proper choice of characterization techniques becomes essential. Among the techniques used to characterize a nanomaterial are scanning electron microscopy (SEM), atomic force microscopy (AFM), spectroscopic analysis, dynamic light scattering (DLS), and X-ray diffraction (XRD), which will be briefly described [53, 72].

1.3.1   Scanning Electron Microscopy

Electron microscopy is widely used in several areas of knowledge such as physics, engineering, materials science, chemistry, biology, and others, for visualizing the structure of materials in the nanometer to micrometer scale, which cannot be observed with the naked eye. SEM made it possible to obtain images, 3D appearances from inside the sample or its surface, of a wide variety of materials, one of the most versatile for analyzing the microstructural characteristics of solid materials.

The principle of SEM is to use a small-diameter electron beam to explore the sample surface, point by point, in successive lines and transmit the detector's signal to a cathode screen whose scanning is perfectly synchronized with that of the incident beam. Through a deflection coil system, the beam can be guided to scan the sample surface using a rectangular mesh. The image signal results from the interaction of the incident beam with the sample surface. The signal collected by the detector is used to modulate the brightness of the monitor, allowing the observation of surface morphology, particle size, with good resolution. When it comes to nanotechnology, SEM becomes an ally for 3D manipulation, assembly, and characterization of nanotubes and nanowires. Thus, during the synthesis and manipulation of nanomaterials, SEM is used as an auxiliary tool to guide the researcher.

1.3.2 Atomic Force Microscopy

In the AFM technique the sample surface is scanned with a probe in order to obtain its topographic image with atomic resolution. The measurement is made by the interaction of probe with the sample surface (forces of attraction and repulsion). This technique is used for non-conductive samples. In the images obtained, it is possible to determine the height and diameter of nanoparticles, if they do not form clusters. AFM allows 3D images with good resolution, but only analyzes the sample surface.

1.3.3 Spectroscopic Analysis

Among the spectroscopic techniques, the most used to characterize a nanomaterial is ultraviolet-visible spectroscopy (UV-Vis) and infrared spectroscopy (IR) and Raman scattering. Absorption spectroscopy in UV-Vis is based on the absorption of ultraviolet/visible radiation (180 to 780 nm) by the sample, which promotes electrons from the fundamental electronic state to the excited state. Thus, the irradiation of light on the diluted solution of the sample provides an absorption spectrum in the wavelength range. It is a versatile technique that can be applied both in the characterization and in the quantification of different types of organic, inorganic, biological materials, etc.

Fourier-transform infrared spectroscopy (FTIR) allows the identification of the elements that make up the analyzed sample, in addition to providing information on the molecular organization and interaction mechanisms. The FTIR principle is the absorption of radiation in the infrared region (above 780 nm). Each type of connection and interaction between elements has a vibrational energy that is related to absorption in the infrared, producing a characteristic spectrum. Thus, it is possible to identify the elements and types of connections present in the analyzed material.

Raman scattering is determined by the inelastic scattering of light that falls on the material. It is a non-destructive, fast, and efficient technique that allows the characterization of organic and inorganic materials in terms of their structure and chemical composition.

1.3.4 Dynamic Light Scattering

The DLS is given by the intensity of the scattered light concerning the particle size distribution, in the submicrometric region and with the latest technology, less than 1 nm. The Brownian movement of particles or molecules in suspension causes the laser light to be spread with different intensities. The analysis of these intensity fluctuations results in the speed of Brownian motion and thus, the particle size using the Stokes–Einstein relationship. These particles can be organic, composed of polymers, carbohydrates, proteins, and surfactants, or inorganic, composed of metals, such as gold or silver nanoparticles, or those formed by transition metal oxides.

1.3.5 X-ray Diffraction

XRD is a characterization technique that consists of the dispersion or spreading of X-rays in all directions by the electrons associated with each atom or ion present in the sample. Thus, it provides information on the arrangement of atoms or ions in the sample analyzed, ranging from the identification and quantification of the crystalline phase, network parameters, and deformation (or micro-deformation) to the estimation of particle size.

The characterization techniques of nanomaterials are shown to be necessary to also assist in toxicity tests, so the knowledge of the physical–chemical and biological properties of the materials is of fundamental importance to obtain information on the risks that it can cause to the environment and human population. The same properties that make nanomaterials so attractive, such as small particle size, varied shape, and high surface area, may also be responsible for harmful effects to living organisms, according to evidence reported by toxicological studies with microorganisms, seaweed, fish, mice, and human cells [73].

1.4 Risks to Human Population and Environment

As they are present in many products and applications of human life from cosmetics, food, tissues for implantation in human beings to materials of environmental treatment, it is expected that their generalized presence allows an easy and quick assessment of possible risks [33]. Vance et al. [74] carried out a survey between 2005 and 2014 on products in common use that contain nanoparticles and observed that 42% were destined to health and fitness care, 37% had metal nanoparticles or metallic oxides, with metallic silver present in 24% of them. In this research, 31% of the evaluated products attributed the use of nanoparticles to their microbial protection effect, while 11% of the products sold the presence of nanoparticles in their composition (silver, TiO_2, gold, nano-organic) as food supplements [74].

In addition, the exposure of workers to nanomaterials is worrying, because very little is known about health effects. These materials can have different properties, and testing of each material is time consuming and costly [16]. Because of this, managing the risks of manufactured nanomaterials requires the ability to accurately and reproducibly measure the physical and chemical properties of these materials that are relevant to their risk assessment [19].

Exposure to both types is currently being investigated and these may enter air, water, and soil media from a range of routes. Physicochemical and biological transformations make nanomaterials potentially highly reactive in both environmental and biological systems, which may alter their fate, dispersion, and toxicity compared with their larger counterparts [34].

The researchers demonstrate that a number of new tools to enable regulatory risk assessment of nanomaterials are now available or near completion [23]. In relation to exposure and risk assessment, relevant issues such as life cycle, bioaccumulation, and administered dose should be considered. Elements to improve the feasibility of carrying out risk assessment in practice include standardized tests, knowledge about in vitro–in vivo comparison, and functional assays. With this information and the need to increase efficiency in risk assessment, it is expected that future perspectives will be presented [55].

1.5 Nanoscale Air Pollutants

Regarding bioaccumulation, nanoparticles should be taken into account, whose origin may be natural, anthropogenic, or incidental (generated as a by-product of anthropogenic processes) [87]. In relation to environmental pollution, a large part of atmospheric pollutants is found on the nanoscale with enormous potential to cause severe damage to the environment and mainly to the human population [54, 79]. This persistent and worldwide air pollution problem tremendously endangers public health due to the existence of poisonous pollutants, which are the mixture of particles, toxic gases, and microorganisms, called particulate matter.

The major components of PM are sulfates, nitrates, sodium chloride, ammonia, black carbon, mineral dust, and water. The most health-damaging particles are those with the diameters of less than 10 μm, which can penetrate directly and lodge deep inside the lungs. Both short-term and long-term exposures to air pollutants have been associated closely with health impacts. $PM_{2.5}$ and PM_{10} are defined of fine particulate matter with an aerodynamic equivalent diameter less than 2.5 μm and 10 μm, respectively [46]. Particularly, fine particles with a diameter less than 2.5 μm in the complex mixture have been acknowledged as the chief hazard because they can easily penetrate deep into human respiratory system, resulting in increased risk of asthma, lung cancer, stroke, heart disease, etc. [39], as illustrated in Fig. 1.3.

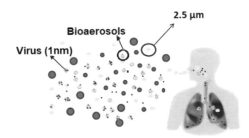

Figure 1.3 Schematic diagram of penetration deep into human respiratory system by fine particulate matter.

The impact can still be reported as physiological impairments, psychological impacts besides the increase in expenses due to the loss of productivity at work and related to increased healthcare expenditure [10, 79, 85].

Microorganisms (e.g. viruses, bacteria, and fungi), fragments of biological particles, and toxins in the air are also classified as PM. Bioaerosols, as they are called, vary in size, from submicroscopic particles (<0.01 µm) to particles larger than 100 µm, which are easily transmitted by the wind and can float for a long time in the atmosphere. If inhaled or adhered to humans, they become a dangerous group of etiological agents for human respiratory and infectious diseases. For this reason, the control of airborne microorganisms is currently an active research field, driven by the growing demand for occupational safety and public health [81]. To protect the environment from the adverse effects of pollution, many countries worldwide enacted legislation to regulate various types of pollution as well as to mitigate the adverse effects of contamination.

1.6   Fibrous Filters and Nanofiltration

Fibrous filters are simple and economical devices to efficiently remove submicrometer particles from gas streams [43]. Currently, nanofiber membranes have been showing superior performance compared to traditional microfiber filtration materials [1, 6, 14, 48]. Membranes derived from electrospun nanofibers with a small diameter, which have high porosity and high specific surface area, are widely used for air filtration. These membrane or filtration systems offer a potential solution for a wide range of environmental issues such as air filtration, water purification, process industries [36, 52, 76, 82]. Membrane processes are regarded as a significant technological progress to ensure the sustainable development of human beings.

Fibrous air filter media are an important kind of air filter that can capture air particles by the synergistic effect of thick physical barriers and adhesion. The fibrous air filter media can achieve high filtration efficiency with relatively low-pressure drop [28]. To remove particles from polluted gas, particles must impact with fibers and be motionless. Thus, sufficient fine fibers, compact stacking structures, high thickness, and remarkable electrostatic effect of filter media are helpful for effective particle capture [39]. Figure 1.4 presents the schematic diagram illustrating the removal of particulate material by a filtering media made of nanofibers.

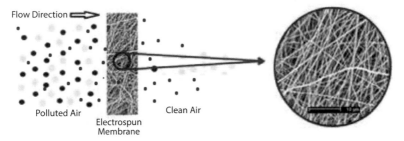

Figure 1.4 Schematic diagram illustrating the filtration by fibrous membranes and SEM image of polymeric membrane.

1.6.1 Theory of Air Filtration

In the last century, the theory of air filtration was already established based on the Brownian movement. Most scholars related the filtration efficiency to the size of the particle to be filtered. Later, they also started to relate filtration to the collection mechanisms that could occur taking into account the tortuosity of the medium when fibrous filter media appeared. The evolution of the filtration theory also sought to explain problems of clogging, and the formation of dendrites that blocked the filtration was related to the concept of pressure drop. What can be concluded from these theories is that filtration can occur in two different ways: with accumulation of material in the filter (non-stationary stage) and without accumulation of material in the filter (stationary stage) [46, 61, 86].

Filtration in fibrous filters (filter media) occurs practically in two stages; however, depending on the filter media or how the cake is formed, it can occur in only one or more stages. In the first stage, the deposition or penetration of particulate material in the fibers occurs, and in this step the pressure drop varies slowly, with interaction between the fibers and the particles (adhesion force), until a thin layer of dust is deposited on the fibers. When this occurs, the pressure drop increases linearly with the filtration time, and the interaction begins to occur between particles (cohesive force). Depending on the filter media, the first stage is practically nonexistent, and the pressure drop increases rapidly according to the second stage of the filtration process [61, 86].

As for airflow with low concentration of particles or high-efficiency filters, the first step is the main one for filtration. Thus, when nanofibers are used to filter nanoparticles in low concentrations, the filtration time will be long enough for the cake to form with consequent variation in pressure drop, being considered stationary. However, if nanofibers are used to filter very high concentrations of nanoparticles and microparticles, the formation of filter cake will be observed in relatively short times [61].

According to the classical filtration theory, there are mainly five mechanisms to catch particles during the first stage, when the interaction between the fiber and the particle occurs until the formation of a thin powder layer on the surface of the filter media. These mechanisms are interception, inertial impact, diffusional effect (Brownian diffusion), gravitational effect, and electrostatic effect. The predominance of one or more mechanisms is related to the size of the particle to be filtered, the flow rate of the gas flow, the characteristics of the fluid, and the diameter of the fibers that make up the filter media as well as its distribution, the latter being amenable to control by the electrospinning technique [87]. Figure 1.5 illustrates the mechanisms to capture particulate by a fibrous layer in the first stage.

In summary, the filtration mechanisms can be explained as follows [47, 61, 86, 87].

1. **Interception effect:** Interception occurs when a particle deviates from the gas flow and collides with the fiber surface under the influence of van der Waals forces.

2. **Inertial deposition:** This mechanism occurs when particles deviate from the flow line due to their larger size. It occurs mainly in particles larger than 0.3–1 µm, especially in the case of a higher gas flow velocity.

3. **Brownian diffusion:** The capture occurs because when moving randomly, deviating from the flow line, the particles collide with the fiber surface leading to deposition in the collector. It occurs especially for particles smaller than

0.1 μm, which exhibit significant Brownian motion, resulting in diffusion and deposition around the surfaces of the fibers.

4. **Electrostatic effect:** It occurs when there are charged particles or when the collector itself is charged. In this way, the particles are attracted, depositing in the collector.
5. **Effect of gravity:** It is caused by the sedimentation of particles in the collector due to the action of gravity. However, due to the small size of the particles, this effect occurs minimally.

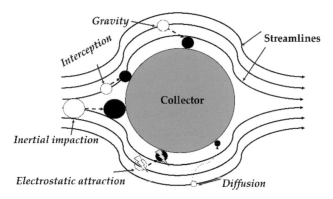

Figure 1.5 Collection mechanisms acting on particles during air filtration, in which the collector is a cylindrical fiber with the axial direction perpendicular to the paper plane. White and black spheres represent particles before and after the influence of the collection mechanism, respectively (adapted from Ref. [25]).

The efficiency of single fiber under every mechanism can be calculated, but the total efficiency of single fiber is not simply the total efficiency under every mechanism, but is the interaction effect of the five mechanisms [46, 86]. Details about equations to estimate fiber mechanisms for capturing particles and efficiency of single fiber can be found in Hinds [25].

It is noteworthy that the collection efficiency of a fibrous filter depends heavily on the single fiber properties [61]. Experimental filtration efficiency can be expressed by the particle concentration of inlet and outlet airflow. The filtration efficiency (η) and resistance (ΔP) are expressed by Eqs. (1.1) and (1.2).

$$\eta = \frac{G_1 - G_2}{G_1} = \frac{Q(N_1 - N_2)}{N_1 Q} = 1 - \frac{N_2}{N_1} \tag{1.1}$$

where G_1 and G_2 are the quantity of particles in inlet and outlet airflow (mg/h); N_1 and N_2 are the particles concentration of inlet and outlet airflow (mg/m^3); and Q is the volumetric flow (m^3/h).

$$\Delta P = \frac{2 * C' v^2 L \rho_g}{\pi d_f^2} (Pa) \tag{1.2}$$

where C' is the resistance coefficient determined experimentally; v is the filtration velocity (m/s); L is the thickness of filtration layer (m); and ρ_g is the gas density (kg/m^3); d_f is the fiber diameter (m) [61].

Several researchers reported experimental filtration data of nanoparticles based on an efficiency curve versus particle diameter. During filtration, there is a predominance of more than one collection mechanism, and in these regions, there is a greater penetration of particles through the filter media, that is, regions of minimal efficiency [47]. According to researchers, the diffusional mechanism is the most active in particles smaller than 0.1 µm, while the inertial and direct interception mechanisms are more active for particles larger than 1.0 µm. However, for particles whose diameter is between 0.1 and 1.0 µm, the minimum efficiency region mentioned earlier can be observed. In this region, due to the greater penetration of particles in the filter, there is a reduction in the general collection efficiency. As the diameter increases (range 0.1 to 0.4 µm), the filtration becomes less efficient, considering that the particles are too large for an effective diffusion effect, but still too small for a significant impact on the interception [6, 8, 86, 87]. This is the expected behavior for filtering nanoparticles by nanofibers for low filtration velocities; however, a small deviation is related to the parameters used by each researcher to measure efficiency, such as fiber and particle diameter, thickness and porosity.

Figure 1.6 presents the collection efficiency of filter media calculated from the theoretical efficiencies of the individual collection mechanisms presented by Hinds [25]. The efficiency curves were calculated for different porosities and sizes of the nanofibers presented in the filter media, considering the collection of NaCl nanoparticles (1–1000 nm) with Boltzmann distribution of charges and at air velocities of 1 and 5 cm/s.

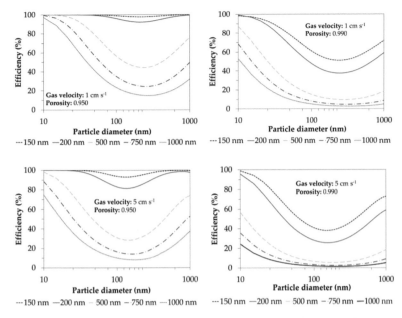

Figure 1.6 Theoretical efficiency of filter media with different median sizes of fibers and porosities collecting NaCl particles (1–1000 nm) at different air velocities.

It is possible to verify in the theoretical curves that the decrease in the size of the fibers increases the collection efficiency for all the ranges of particle size, porosity, and air velocity evaluated, especially in the range of particle size around 100 nm, in which the efficiency curves present minimum values [2]. These results justify the use of nanofibers in air-filtration applications. In addition, it is observed that the increase in porosity, which means the increase in voids over the thickness of the filter media, decreases the collection efficiency in the particle size range evaluated [6]. Therefore, the porosity of the filter media has to be controlled in the manufacturing of the filter media as well, not only to provide low energy consumption by means of low-pressure drop, but also to provide high collection efficiencies. Finally, it must be noticed that the increase in the air velocity decreases the collection efficiency, which is strongly associated to the decrease in the efficiency of the diffusion mechanism for this particle size range [28].

1.6.2   Air-Filtration Applications: Nanofibers

For this reason, nanofibers have become the major interest of researchers because of their desirable properties for filtration, such as mechanical strength, elasticity, porosity, charged surface area, among others [51]. In contrast to the micro-sized fibers with an average diameter from several to tens of micrometers, the nanofibrous air filters with an average diameter below 1.0 μm exhibit higher removal efficiency toward ultrafine PM [26].

Regarding the filter media, it is important to guarantee high filtration efficiency, low-pressure drop, and low operating cost, and these are related to porosity, permeability, fiber diameter, thickness, and surface speed of filtration. These are the factors that should guide the production process of the filter media so that the removal of particles is efficient and with low-pressure drop. A wide variety of polymers have been used in the preparation of these nanofibers such as polyamide, polyimide, polystyrene, polyacrylonitrile, polyacid acetic, polyvinyl acetate [22, 49], and it has shown satisfactory results, high collection efficiency, and low-pressure drop.

Bortolassi et al. [6] studied the efficient removal of nanoparticles and the bactericidal action with nanofibers for applications in air filtration. The nanofibers (300 nm) were produced under a microfiber substrate and achieved collection efficiency close to 100% for particles whose diameter varied from 9 to 300 nm with quality factor equal to 0.06 Pa^{-1} at 5 cm/s of air. According to the authors, to explain the curves behavior of filtration efficiency versus particle size, three main mechanisms were used: interception, inertial impaction, and diffusion. They also stressed the importance of nanofiber layers to increase collection efficiency, as the filtration efficiency of the common fibrous filter increases with the increase in the deposition of layers, directly proportional to the air pressure drop [6].

Li et al. [38] made electrospinning polyimide nanofibers (P84) (200–500 nm) on needle punched aramid felt pretreated with electrospray in order to enhance the adhesion between the layers of fibers. The authors reported an increase in the initial collection efficiency from 48.56% (for felt without nanofibers) to 94.83%

(with 120 min of electrospinning), with quality factors subsequently equal to 0.0175 and 0.0214 Pa^{-1}, when tested using NaCl particles (0.3–10 µm) at 10 cm/s of air. The quicker cake formation due to the presence of the nanofibers diminished the occurrence of depth filtration in relation to the samples without the nanofibers. This phenomenon prevents the clogging of the filter media, diminishing the operational costs involving the cleaning and the exchanging of the filter media [38].

Zhang et al. developed polyimide nanofibers (300 nm) to remove PM$_{2.5}$ for high-temperature uses. In tests whose thermal stability was required, they reached an efficiency of 99.5% at high temperatures. They also reported that the temperature variation (from 25 to 370°C) did not change the diameter and morphology of the nanofibers, ensuring thermal stability. In this way, it can be concluded that the high collection efficiency reached the HEPA filter standard defined as filters with filtration efficiency at 99.97% for 0.3 µm in air particles. A quality factor of 0.1 Pa^{-1} at 0.2 m/s of air was observed at low-pressure drops. These values were due to the small diameter of the nanofibers and the small thickness of the manufactured filter media (0.01 to 0.1 mm). The authors used the effect of sliding particles under the filter media to explain this behavior, because when the diameter of the nanofibers is comparable to the average free path of air molecules (66 nm under normal conditions), the gas velocity is different from zero on the fiber surface due to the "slip" effect. Because of the "slip" effect, the drag force of the nanofibers in the airflow decreases, leading to reduction in pressure drop [84].

The association of micro- and nanofibers was evaluated by Deng et al. [13] for making filter media from polypropylene and polystyrene membranes. It was reported that the microscale fibers (3 µm) acted as support for the nanofibers (300 nm), improving the permeability of the medium, while the nanofibers, due to the greater area/volume ratio, improved the filtration performance. The filter media showed filtration efficiency close to 99.87% for particles whose average diameter was 0.26 µm, a low-pressure drop of 37.73 Pa, and a quality factor of 0.18 Pa^{-1}. This work provides a new strategy for the development of materials applied to air filtration [13].

1.7 Final Considerations and Perspectives

Owing to these advantageous features and intriguing characteristics of the nanofibers, their production has attracted much attention for the preparation of functional air filters [40]. A variety of processing techniques such as drawing, electrospinning, co-electrospinning, centrifugal spinning, blow spinning, force spinning, melt spinning, melt blow spinning, dry spinning, wet spinning, gel spinning, template synthesis, phase separation, self-assembly [27, 51], and atomic layer deposition [3, 56] have been used to prepare polymeric nanofibers in recent years. Among the techniques, the electrospinning process stands out, which consists of producing nanofibers through the action of electrical forces [64]. The main characteristics of the nanofiber production methods will be discussed in Chapter 2.

Therefore, due to the widespread dissemination of nanoscience concepts and the rapid advance of nanotechnology, it was possible to develop nanomaterials bringing new technologies and new materials with the most diverse applications, which are helping society, health, environment, and industry. However, nanoparticles in the air can cause serious damage to health and they must be captured. Thus, nanofibers have been standing out in air filtration, with the manufacture of intelligent high collection efficiency membranes for nanoparticles compared to the currently commercialized filters, being a very promising technology. The operational performance of these nanofibers has stood out in the mitigation of air pollution, added to the unique characteristics of nanofibers such as high surface area, high porosity, adaptations of functionality, hybridization of morphologies, in addition to the low cost of material. Due to all of these, the study, knowledge, and application of nanofibers in the most diverse sectors, above all, in the control of air pollution, have increased in recent years, which translates into great potential for application, driving more and more research on applied nanotechnology.

References

1. Al-Attabi, R., Dumée, L. F., Schütz, J. A., and Morsi, Y. (2018). Pore engineering towards highly efficient electrospun nanofibrous

membranes for aerosol particle removal, *Sci. Total Environ.*, **625**, pp. 706–715.

2. Balgis, R., Kartikowati, C. W., Ogi, T., Gradon, L., Bao, L., Seki, K., and Okuyama, K. (2015). Synthesis and evaluation of straight and bead-free nano fibers for improved aerosol filtration, *Chem. Eng. Sci.*, **137**, pp. 947–954.

3. Bechelany, M., Drobek, M., Vallicari, C., Abou Chaaya, A., Julbe, A., and Miele, P. (2015). Highly crystalline MOF-based materials grown on electrospun nanofibers, *Nanoscale,* **7**, pp. 5794–5802.

4. Binnig, G. and Rohrer, H. (1984). Scanning tunneling microscopy, *Phys. B C,* **127**, pp. 37–45.

5. Borm, P. J. A., Robbins, D., Haubold, S., Kuhlbusch, T., Fissan, H., Donaldson, K., Schins, R., Stone, V., Kreyling, W., Lademann, J., Krutmann, J., Warheit, D. B., and Oberdorster, E. (2006). The potential risks of nanomaterials: A review carried out for ECETOC, *Part. Fibre Toxicol.*, **3**, pp. 1–35.

6. Bortolassi, A. C. C., Nagarajan, S., de Araújo Lima, B., Guerra, V. G., Aguiar, M. L., Huon, V., Soussan, L., Cornu, D., Miele, P., and Bechelany, M. (2019). Efficient nanoparticles removal and bactericidal action of electrospun nanofibers membranes for air filtration, *Mater. Sci. Eng. C,* **102**, pp. 718–729.

7. Bratovcic, A. (2019). Different applications of nanomaterials and their impact on the environment, *Int. J. Mater. Sci. Eng.*, **5**, pp. 1–7.

8. Cao, M., Gu, F., Rao, C., Fu, J., and Zhao, P. (2019). Improving the electrospinning process of fabricating nanofibrous membranes to filter PM2, *Sci. Total Environ.*, **666**, pp. 1011–1021.

9. Christian, P., Von Der Kammer, F., Baalousha, M., and Hofmann, T. (2008). Nanoparticles: Structure, properties, preparation and behaviour in environmental media, *Ecotoxicology*, **17**, pp. 326–343.

10. Cohen, A. J., Brauer, M., Burnett, R., Anderson, H. R., Frostad, J., Estep, K., Balakrishnan, K., Brunekreef, B., Morawska, L., Iii, C. A. P., Shin, H., Straif, K., Shaddick, G., Thomas, M., Dingenen, R. Van, Donkelaar, A. Van, Vos, T., Murray, C. J. L., and Forouzanfar, M. H. (2015). Estimates and 25-year trends of the global burden of disease attributable to ambient air pollution: An analysis of data from the Global Burden of Diseases Study 2015, *Lancet*, **389**, pp. 1907–1918.

11. David, R. M. (2013). Measuring engineered nanomaterials in the environment: A consortium view of how to address the problem, *Environ. Eng. Sci.*, **30**, pp. 97–100.

12. De Marchi, L., Coppola, F., Soares, A. M. V. M., Pretti, C., Monserrat, J. M., Torre, C. della, and Freitas, R. (2019). Engineered nanomaterials: From their properties and applications, to their toxicity towards marine bivalves in a changing environment, *Environ. Res.*, **178**, pp. 108683.

13. Deng, N., He, H., Yan, J., Zhao, Y., Ben, E., and Liu, Y. (2019). One-step melt-blowing of multi-scale micro/nanofabric membrane for advanced air-filtration, *Polymer*, **165**, pp. 174–179.

14. Di, X., Zhang, W., Zang, D., Liu, F., Wang, Y., and Wang, C. (2016). A novel method for the fabrication of superhydrophobic nylon net, *Chem. Eng. J.*, **306**, pp. 53–59.

15. Elessawy, N. A., Elnouby, M., Gouda, M. H., Hamad, H. A., Taha, N. A., Gouda, M., and Mohy Eldin, M. S. (2020). Ciprofloxacin removal using magnetic fullerene nanocomposite obtained from sustainable PET bottle wastes: Adsorption process optimization, kinetics, isotherm, regeneration and recycling studies, *Chemosphere*, **239**, pp. 124728.

16. Elisabeth, N., Skaug, V., Mohr, B., Verbeek, J., and Zienolddiny, S. (2018). Criteria for grouping of manufactured nanomaterials to facilitate hazard and risk assessment: A systematic review of expert opinions, *Regul. Toxicol. Pharmacol.*, **95**, pp. 270–279.

17. Englert, B. C. (2007). Nanomaterials and the environment: Uses, methods and measurement, *J. Environ. Monit.*, **9**, pp. 1154–1161.

18. Farré, M., Sanchís, J., and Barceló, D. (2011). Analysis and assessment of the occurrence, the fate and the behavior of nanomaterials in the environment, *Trends Anal. Chem.*, **30**, pp. 517–527.

19. Gao, X. and Lowry, G. V. (2018). Progress towards standardized and validated characterizations for measuring physicochemical properties of manufactured nanomaterials relevant to nano health and safety risks, *NanoImpact*, **9**, pp. 14–30.

20. Gleiter, H. (2000). Nanostructured materials: Basic concepts and microstructure, *Acta Mater.*, **48**, pp. 1–29.

21. Greiner, A., Wendorff, J. H., Yarin, A. L., and Zussman, E. (2006). Biohybrid nanosystems with polymer nanofibers and nanotubes, *Appl. Microbiol. Biotechnol.*, **71**, pp. 387–393.

22. Guibo, Y., Qing, Z., Yahong, Z., Yin, Y., and Yumin, Y. (2012). The electrospun polyamide 6 nanofiber membranes used as high efficiency filter materials: Filtration potential, thermal treatment, and their continuous production, *J. Appl. Polym. Sci.*, **128**, pp. 1061–1069.

23. Günter, K. and Sayre, P. G. (2017). Reliability of methods and data for regulatory assessment of nanomaterial risks, *NanoImpact*, **7**, pp. 66–74.

24. Hey, T. (1999). Richard Feynman and computation, *Contemp. Phys.*, **40**, pp. 257–265.

25. Hinds, C. W. (1998). *Aerosol Technology: Properties, Behaviour, and Measurement of Airborne Particles*, 2nd Ed. (John Wiley, USA).

26. Huang, J. J., Tian, Y., Wang, R., Tian, M., and Liao, Y. (2020). Fabrication of bead-on-string polyacrylonitrile nano fibrous air filters with superior filtration efficiency and ultralow pressure drop, *Sep. Purif. Technol.*, **237**, pp. 116377.

27. Huang, Z., Zhang, Y., Kotaki, M., and Ramakrishna, S. (2003). A review on polymer nanofibers by electrospinning and their applications in nanocomposites, *Compos. Sci. Technol.*, **63**, pp. 2223–2253.

28. Huang, Z. X., Liu, X., Zhang, X., Wong, S. C., Chase, G. G., Qu, J. P., and Baji, A. (2017). Electrospun polyvinylidene fluoride containing nanoscale graphite platelets as electret membrane and its application in air filtration under extreme environment, *Polymer*, **131**, pp. 143–150.

29. International Organization for Standardization (ISO). ISO/TC 229: *Nanotechnologies.*

30. International Organization for Standardization (ISO). ISO/TS 80004: *Nanotechnologies.*

31. International Organization for Standardization (ISO). ISO/TS 27687: *Nanotechnologies.*

32. Jeevanandam, J., Barhoum, A., Chan, Y. S., Dufresne, A., and Danquah, M. K. (2018). Review on nanoparticles and nanostructured materials: History, sources, toxicity and regulations, *Beilstein J. Nanotechnol.*, **9**, pp. 1050–1074.

33. Kühnel, D., Nickel, C., Hellack, B., Zalm, E. Van Der, Kussatz, C., Herrchen, M., Meisterjahn, B., and Hund-rinke, K. (2019). Closing gaps for environmental risk screening of engineered nanomaterials, *NanoImpact*, **15**, pp. 100173.

34. Kumar, P., Kumar, A., Fernandes, T., and Ayoko, G. A. (2014). Nanomaterials and the environment, *J. Nanomater.*, **2014**, pp. 1–4.

35. Kumar, V., Lee, Y. S., Shin, J. W., Kim, K. H., Kukkar, D., and Fai Tsang, Y. (2020). Potential applications of graphene-based nanomaterials as adsorbent for removal of volatile organic compounds, *Environ. Int.*, **135**, pp. 105356.

36. Ladd, M. R., Hill, T. K., Yoo, J. J., and Lee, S. J. (2011). Electrospun nanofibers in tissue engineering, in *Nanofibers: Production, Properties and Functional Applications* (IntechOpen), pp. 347–372.

37. Li, C., Huang, G., Cheng, G., Zheng, M., and Zhou, N. (2019). Nanomaterials in the environment: Research hotspots and trends, *Int. J. Environ. Res. Public Health*, **16**, pp. 5138.

38. Li, M., Feng, Y., Wang, K., Yong, W. F., Yu, L., and Chung, T. (2017). Novel hollow fiber air filters for the removal of ultrafine particles in $PM_{2.5}$ with repetitive usage capability, *Environ. Sci. Technol.*, **51**, pp. 10041–10049.

39. Li, Y., Yin, X., Yu, J., and Ding, B. (2019). Electrospun nanofibers for high-performance air filtration, *Compos. Commun.*, **15**, pp. 6–19.

40. Liu, F., Li, M., Shao, W., Yue, W., Hu, B., Weng, K., Chen, Y., Liao, X., and He, J. (2019). Preparation of a polyurethane electret nanofiber membrane and its air-filtration performance, *J. Colloid Interface Sci.*, **557**, pp. 318–327.

41. Liu, H., Zhang, S., Liu, L., Yu, J., and Ding, B. (2020). High-performance PM 0.3 air filters using self-polarized electret nanofiber/nets, *Adv. Funct. Mater.*, **30**, pp. 1909554.

42. Liu, J., Jiang, T., Li, X., and Wang, Z. L. (2019). Triboelectric filtering for air purification, *Nanotechnology*, **30**, pp. 29001.

43. Liu, Y., Park, M., Ding, B., Kim, J., El-Newehy, M., Al-Deyab, S. S., and Kim, H. Y. (2015). Facile electrospun polyacrylonitrile/poly(acrylic acid) nanofibrous membranes for high efficiency particulate air filtration, *Fibers Polym.*, **16**, pp. 629–633.

44. Louro, H., Borges, T., and Silva, M. J. (2013). Manufactured nanomaterials: New challenges for public health, *Rev. Port. Saude Publica*, **31**, pp. 145–157.

45. Lozano, L. A., Faroldi, B. M. C., Ulla, A., and Zamaro, J. M. (2020). Metal-organic framework-based sustainable nanocatalysts for CO oxidation, *Nanomaterials*, **10**, pp. 165.

46. Lv, D., Zhu, M., Jiang, Z., Jiang, S., Zhang, Q., and Xiong, R. (2018). Green electrospun nanofibers and their application in air filtration, *Macromol. Mater. Eng.*, **303**, pp. 1800336.

47. Lv, M., Geng, J., Kou, X., Xin, Z., and Yang, D. (2018). Engineering nanomaterials-based biosensors for food safety detection, *Biosens. Bioelectron.*, **106**, pp. 122–128.

48. Matulevicius, J., Kliucininkas, L., Prasauskas, T., and Buivydiene, D. (2016). The comparative study of aerosol filtration by electrospun polyamide, polyvinyl acetate, polyacrylonitrile and cellulose acetate nanofiber media, *J. Aerosol Sci.*, **92**, pp. 27–37.

49. Matulevicius, J., Kliucininkas, L., Prasauskas, T., Buivydiene, D., and Martuzevicius, D. (2016). The comparative study of aerosol filtration by electrospun polyamide, polyvinyl acetate, polyacrylonitrile and cellulose acetate nanofiber media, *J. Aerosol Sci.*, **92**, pp. 27–37.

50. Mendonça, M. C. P., Rizoli, C., Ávila, D. S., Amorim, M. J. B., and de Jesus, M. B. (2017). Nanomaterials in the environment: Perspectives on in Vivo terrestrial toxicity testing, *Front. Environ. Sci.*, **5**, pp. 71.

51. Mercante, L. A., Scagion, V. P., Migliorini, F. L., Mattoso, L. H. C., and Correa, D. S. (2017). Trends in analytical chemistry electrospinning-based (bio) sensors for food and agricultural applications: A review, *Trends Anal. Chem.*, **91**, pp. 91–103.

52. Mishra, R. K., Mishra, P., Verma, K., Mondal, A., Chaudhary, R. G., Abolhasani, M. M., and Loganathan, S. (2019). Electrospinning production of nanofibrous membranes, *Environmental Chemistry Letters*, **17**, pp. 767–800.

53. Mourdikoudis, S., Pallares, R. M., and Thanh, N. T. K. (2018). Characterization techniques for nanoparticles: Comparison and complementarity upon studying nanoparticle properties, *Nanoscale*, **10**, pp. 12871–12934.

54. Ok, V., Han, Y., and Ck, J. (2018). Air pollution and environmental injustice: Are the socially deprived exposed to more $PM_{2.5}$ pollution in Hong Kong? *Environ. Sci. Policy*, **80**, pp. 53–61.

55. Oomen, A. G., Günter, K., Bleeker, E. A. J., Broekhuizen, F. Van, Sips, A., Dekkers, S., Wijnhoven, S. W. P., and Sayre, P. G. (2018). Risk assessment frameworks for nanomaterials: Scope, link to regulations, applicability, and outline for future directions in view of needed increase in efficiency, *NanoImpact*, **9**, pp. 1–13.

56. Palmstrom, A. F., Santra, P. K., and Bent, S. F. (2015). Atomic layer deposition in nanostructured photovoltaics: Tuning optical, electronic and surface properties, *Nanoscale*, **7**, pp. 12266–12283.

57. Peng, Z., Liu, X., Zhang, W., Zeng, Z., Liu, Z., Zhang, C., Liu, Y., Shao, B., Liang, Q., Tang, W., and Yuan, X. (2020). Advances in the application, toxicity and degradation of carbon nanomaterials in environment: A review, *Environ. Int.*, **134**, pp. 105298.

58. Petkova-Georgieva, S. (2019). Impact of nanomaterials usage on human health and nature environment, *J. Environ. Prot. Ecol.*, **20**, pp. 2093–2102.

59. Petronella, F., Truppi, A., Dell'Edera, M., Agostiano, A., Curri, M. L., and Comparelli, R. (2019). Scalable synthesis of mesoporous TiO_2 for

environmental photocatalytic applications, *Materials (Basel).*, **12**, pp. 1853.

60. Promphet, N., Rattanarat, P., Chailapakul, O., Rangkupan, R., and Rodthongkum, N. (2015). An electrochemical sensor based on graphene/polyaniline/polystyrene nanoporous fiber modified electrode for simultaneous determination of lead and cadmium, *Sens. Actuators B Chem.*, **207**, pp. 526–534.

61. Qin, X. and Wang, S. (2006). Filtration properties of electrospinning nanofibers, *J. Appl. Polym.*, 102, pp. 1285–1290.

62. Qiu, L., Zhu, N., Feng, Y., Michaelides, E. E., Żyła, G., Jing, D., Zhang, X., Norris, P. M., Markides, C. N., and Mahian, O. (2020). A review of recent advances in thermophysical properties at the nanoscale: From solid state to colloids, *Phys. Rep.*, **843**, pp. 1–81.

63. Qureshi, U. A., Khatri, Z., Ahmed, F., Khatri, M., and Kim, I. (2017). Electrospun zein nano fiber as a green and recyclable adsorbent for the removal of reactive black 5 from the aqueous phase, *ACS Sustain. Chem. Eng.*, **5**, pp. 4340–4351.

64. Ramakrishnan, R., Ramakrishnan, P., Ranganathan, B., Tan, C., Sridhar, T. M., and Gimbun, J. (2019). Effect of humidity on formation of electrospun polycaprolactone nanofiber embedded with curcumin using needdleless electrospinning, *Mater. Today Proc.*, **19**, pp. 1241–1246.

65. Sabra, S., Ragab, D. M., Agwa, M. M., and Rohani, S. (2020). Recent advances in electrospun nanofibers for some biomedical applications, *Eur. J. Pharm. Sci.*, **144**, pp. 105224.

66. San-Miguel, A. (2006). Nanomaterials under high-pressure, *Chem. Soc. Rev.*, **35**, pp. 876–889.

67. Selvam, A. K. and Nallathambi, G. (2015). Polyacrylonitrile/silver nanoparticle electrospun nanocomposite matrix for bacterial filtration, *Fibers Polym.*, **16**, pp. 1327–1335.

68. Shanmuganathan, R., Edison, T. N. J. I., LewisOscar, F., Kumar, P., Shanmugam, S., and Pugazhendhi, A. (2019). Chitosan nanopolymers: An overview of drug delivery against cancer, *Int. J. Biol. Macromol.*, **130**, pp. 727–736.

69. Shehzad, K., Xu, Y., Gao, C., and Duan, X. (2016). Three-dimensional macro-structures of two-dimensional nanomaterials, *Chem. Soc. Rev.*, **45**, pp. 5541–5588.

70. Tahir, M. B., Nawaz, T., Nabi, G., Sagir, M., Shehzad, M. A., Yasmin, A., Hussain, S., Bhatti, M. P., Ahmed, A., and Gilani, S. S. A. (2020). Recent

advances on photocatalytic nanomaterials for hydrogen energy evolution in sustainable environment, *Int. J. Environ. Anal. Chem.*, pp. 1–19.

71. Tasciotti, E., Cabrera, F. J., Evangelopoulos, M., Martinez, J. O., Thekkedath, U. R., Kloc, M., Ghobrial, R. M., Li, X. C., Grattoni, A., and Ferrari, M. (2016). The emerging role of nanotechnology in cell and organ transplantation, *Transplantation*, **100**, pp. 1629–1638.

72. Titus, D., James Jebaseelan Samuel, E., and Roopan, S. M. (2019). Nanoparticle characterization techniques, in *Green Synthesis, Characterization and Applications of Nanoparticles* (Elsevier Inc.), pp. 303–319.

73. Tong, Z., Bischoff, M., Nies, L., and Applegate, B. (2007). Impact of fullerene (C60) on a soil microbial community, *Environ. Sci. Technol.*, **41**, pp. 2985–2991.

74. Vance, M. E., Kuiken, T., Vejerano, E. P., Mcginnis, S. P., Jr, M. F. H., Rejeski, D., and Hull, M. S. (2015). Nanotechnology in the real world: Redeveloping the nanomaterial consumer products inventory, *Beilstein J. Nanotechnol.*, **6**, pp. 1769–1780.

75. Wang, G. (2018). *Nanotechnology: The New Features*. arXiv.org.

76. Wang, N., Raza, A., Si, Y., Yu, J., Sun, G., and Ding, B. (2013). Tortuously structured polyvinyl chloride/polyurethane fibrous membranes for high-efficiency fine particulate filtration, *J. Colloid Interface Sci.*, **398**, pp. 240–246.

77. Waychunas, G. A. and Zhang, H. (2008). Structure, chemistry, and properties of mineral nanoparticles, *Elements*, **4**, pp. 381–387.

78. Whatmore, R. W. (1999). Ferroelectrics, microsystems and nanotechnology, *Ferroelectrics*, **225**, pp. 179–192.

79. Yang, J. and Zhang, B. (2018). Air pollution and healthcare expenditure: Implication for the bene fit of air pollution control in China, *Environ. Int.*, **120**, pp. 443–455.

80. Yu, X., Tang, Z., Sun, D., Ouyang, L., and Zhu, M. (2017). Recent advances and remaining challenges of nanostructured materials for hydrogen storage applications, *Prog. Mater. Sci.*, **88**, pp. 1–48.

81. Yun, D., Joon, K., Kang, J., Jeong, E., Jung, S., and Uk, B. (2018). Washable antimicrobial polyester/aluminum air filter with a high capture efficiency and low pressure drop. *J. Hazard. Mater.*, **351**, pp. 29–37.

82. Zander, N. E., Gillan, M., and Sweetser, D. (2016). Recycled PET nanofibers for water filtration applications, *Materials*, **9**, pp. 247.

83. Zarbin, A. J. G. (2007). Química de (nano)materiais, *Quim. Nova*, **30**, pp. 1469–1479.

84. Zhang, R., Liu, C., Hsu, P., Zhang, C., Liu, N., and Zhang, J. (2016). Nanofiber air filters with high-temperature stability for efficient PM 2.5 removal from the pollution sources, *Nano Lett.*, **16**, pp. 3642–3649.

85. Zhong, L., Wang, T., Liu, L., Du, W., and Wang, S. (2018). Ultra-fine SiO_2 nanofilament-based PMIA: A double network membrane for efficient filtration of PM particles, *Sep. Purif. Technol.*, **202**, pp. 357–364.

86. Zhu, C., Lin, C., and Shun, C. (2000). Inertial impaction-dominated fibrous filtration with rectangular or cylindrical fibers, *Powder Techno.*, **112**, pp. 149–162.

87. Zhu, M., Han, J., Wang, F., Shao, W., and Xiong, R. (2017). Electrospun nanofibers membranes for effective air filtration, *Macromol. Mater. Eng.*, **302**, pp. 1600353.

Complementary References

88. Hussain, C. M. (2020). *Handbook of Functionalized Nanomaterials for Industrial Applications*, 1st Ed. (Elsevier).

89. Hussain, C. M. (2020). *The ESLI Handbook of Nanotechnology: Risk Safety, ESLI and Commercialization*, 1st Ed. (Wiley).

90. Nanotechnology Guidance Documents. fda.gov/science-research/nanotechnology-programs-fda/nanotechnology-guidance-documents

91. The Nanodatabase. nanodb.dk/en

Chapter 2

Nanofiber Production Techniques Applied to Filtration Processes

Alessandro Estarque de Oliveira, Daniela Patrícia Freire Bonfim, Ana Isabela Pianowski Salussoglia, Gabriela Brunosi Medeiros, Vádila Giovana Guerra, and Mônica Lopes Aguiar
Graduate Program of Chemical Engineering, Federal University of São Carlos, Rodovia Washington Luís, km 235, PO Box 676, Zip Code 13560-970, São Carlos, SP, Brazil
mlaguiar@ufscar.br

Nanoscience and nanotechnology gave rise to the nanofibers, which in recent years have been attracting the attention of several research centers and the industrial sector and showing increased demand for these nanomaterials. Due to their specific characteristics, such as a high surface-area-to-volume ratio, nanofibers can be applied in several operational fields, such as air and water filtration, reaction catalysis and energy production and storage, such as solar panels and batteries. In addition, they are being used in biomedicine, assisting in the reconstruction of tissues and in the transport of

Environmental, Ethical, and Economical Issues of Nanotechnology
Edited by Chaudhery Mustansar Hussain and Gustavo Marques da Costa
Copyright © 2022 Jenny Stanford Publishing Pte. Ltd.
ISBN 978-981-4877-76-3 (Hardcover), 978-1-003-26185-8 (eBook)
www.jennystanford.com

medicines within our body, in the manufacturing of tissues, or also in structures that require high mechanical resistance and reduced weight, photovoltaic devices, high-performance electrodes, among others.

In air filtration, the main characteristics of filter media containing nanofibers are a high surface-to-volume ratio, an increase in pressure drop, easy removal of the filter cake (particles are trapped in the nanofiber layer), and high filtration performance. Thus, this chapter aims to present an updated description of the nanofibers science and technology most used in filtration processes, nanofiber production techniques, and chemical and physical characterization of the nanofiber filter media. Performance of nanofiber filter media (membranes) in separation, microfiltration, and nanofiltration processes will also be presented. This chapter will highlight the changes in the surface of filter media used in the industrial sector and artificial ventilation, adding nanofibers to increase the collection efficiency for nanoparticles, reduce pressure drop, operation and maintenance costs.

Although focused on air filtration, this chapter will cover several other applications in order to highlight the wide applicability of the described technologies used to manufacture nanofibers and nanostructures.

2.1 Production Techniques of Nanofibers and Nanostructures

The follow production techniques comprise drawing, electrospinning, co-electrospinning, centrifugal spinning, blow spinning, force spinning, melt spinning, melt blow spinning, dry spinning, wet spinning, gel spinning, template synthesis, phase separation, and self-assembly, highlighting different applications and advances in these technologies nowadays. However, distinctions between these techniques are not totally defined many times, since different systems and configurations have been proposed over the years that merge their principles and advantages in order to create enhanced technologies according to the desired applications, which can be found out in the next pages.

2.1.1 Drawing

The first use of drawing to produce fibers in the nanometric scale has been attributed to Ondarçuhu and Joaquim [57]. The authors used a micropipette to pull the nanofibers from a droplet of a citrate solution disposed in a surface, similar to the representation of Fig. 2.1, obtaining fibers from 2 to 100 nm of diameter and with lengths up to millimetric sizes, which could be manipulated by using an atomic force microscope (AFM) tip. According to the authors, only materials with viscoelastic behavior and enough cohesiveness can be used to produce nanofibers by this method since it promotes strong deformations and stresses during the pulling. This latter step, by the way, is accompanied by the solidification of the material, which can be promoted by the cooling of the molten material or the evaporation of the solvent in the case of solutions.

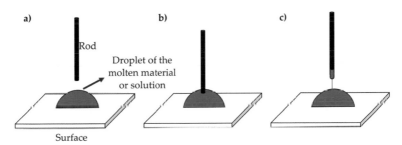

Figure 2.1 Schematics of drawing to produce nanofibers: (a, b) A rod or capillary tube contacts a droplet of the molten material or the solution lying in a surface; (c) and the constant pulling of the object with a controlled rate produce the nanofibers.

The technology of drawing to produce nanofibers evolved since then, by using nanorobotic fiber fabrication to construct 1D, 2D, and 3D structures made of nanofibers [55], and more recently by applying a supersonic laser of CO_2 in a micrometric fiber inside a high-temperature vacuum chamber to produce nanofibers of poly(ethylene terephthalate) (PET) [82, 83], ethylene tetrafluoroethylene (ETFE) [80], Nylon-66 [24], and poly(p-phenylene sulfide) (PPS) [36], for example.

2.1.2 Electrospinning

Electrospinning has been extensively and successfully used in the last decades to produce nanofibers of several materials and for a series of applications. In accordance with the scheme presented in Fig. 2.2, this method consists of passing a polymeric fluid (a solution, in general), through a capillary tube or needle connected to a high-voltage supply, which is also connected to a rotary or static collector located a certain distance away from the capillary tube. As the polymeric fluid flows through the tube (by an infusion pump, for example), an electric potential difference greater than a critical value is applied by the voltage supply and creates an electric field between collector and tube in a way that the drop of fluid exiting the tube tip deforms in a shape called Taylor cone, from which a jet of the fluid releases, bends, and stretches in accordance with the balance between the electrical and viscous forces and the surface tension of the fluid. Thin nanofibers reach the collector as a result of these phenomena and the subsequent evaporation of the solvent in the case of polymeric solutions [65, 66, 105].

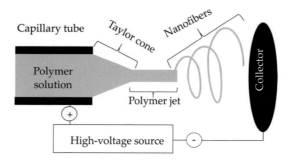

Figure 2.2   Sketch of the electrospinning process.

The great applicability of nanofibers produced by electrospinning ranges from air filtration [29, 40, 46, 100, 101], 3D cellular culture [49], magnetic resonance images [112], photocatalysis [102], detection of gases and volatile organic compounds [76], drug release [27], liquid filtration [16], and electrochemistry [108].

Different works in the literature have reported the use of electrospinning to produce nanofibers of polymers containing nanoparticles of metals or oxides that provide bactericidal properties to the electrospun membranes, essentially silver [1, 73], titanium, and zinc oxides [73]. In the case of nanofibers for air-filtration applications, Bortolassi et al. [8, 9] manufactured poly(acrylonitrile) (PAN) nanofibers containing different nanoparticle contents (TiO_2, ZnO, and Ag) that provided collection efficiencies of NaCl nanoparticles (9–500 nm) of almost 100% with quality factors of approximately 0.05 [9] and 0.06 Pa^{-1} and excellent antibacterial activity for 10^4 [9] and 10^8 CFU/mL of *Escherichia coli* bacteria. Due to these characteristics, these membranes (filter media) can be used in artificial ventilation systems (air conditioning) in hospital environments and in personal protective equipment (PPE) masks, because they can retain even viruses due to their high efficiency. In the retention of nanoaerosols, one that viruses have a size between 20 and 400 nm, bacteria have a size between 0.2 and 2.0 μm, and fungal spores, 2.0 to 8.0 μm [33, 79, 113]. Figure 2.3 presents SEM images of the electrospun PAN nanofibers.

Figure 2.3   NaCl particles adhered to PAN nanofibers used as filter media [7].

Although synthetic polymers have been widely used in the electrospinning process, nanofibers of cellulose compounds have also been manufactured by this technique for different applications. Cellulose acetate nanofibers were produced with silver nanoparticles by electrospinning to obtain antimicrobial activity [75]. Sulfisoxazole/cyclodextrin was incorporated in electrospun hydroxypropyl cellulose nanofibers for drug-delivery applications [5]. In addition, cellulose compounds have been incorporated to electrospun nanofibers with different methodologies to take advantage of the chemical, thermal, and physical resistance of the cellulose, as the incorporation of cellulose nanocrystals that enhanced the mechanical resistance of electrospun poly(lactic acid) fibers [11] and the coating of poly(vinylidene fluoride)-co-hexafluoropropylene (PVDF–HFP) nanofibers with a cellulose-ionic liquid solution that enhanced the mechanical resistance of the fibers in order to be used in oil/water separation [3].

2.1.3   Co-electrospinning and Solution Blow Spinning

In the scope of electrospinning, the so-called co-electrospinning or coaxial electrospinning is a technology in which fibers of two different polymers are electrospun concentrically in order to obtain core–shell structures that take advantage of both materials properties. For example, Rahimi and Mokhtari [63] optimized the operating conditions of co-electrospinning of hexadecane nanofibers (core) in polyurethane fibers (shell), to use the phase change property of the first and the mechanical resistance of the latter to protective clothing applications (i.e., breathable clothes or that allow transpiration while prevent penetration of rain or liquid to keep the body dry and warm). In the work of Zhou et al. [112], polycaprolactone (PCL) and PCL-polysiloxane-based surfactant (PSi) were co-electrospun as shell fibers of poly(ethylene oxide) (PEO) nanofibers (the core) to be applied in magnetic resonance images. Yu et al. [107] used dodecylbenzene sulfonate (SDBS) solutions as sheath fluids to co-electrospinning poly(acrylonitrile) (PAN) core nanofibers down to an average diameter of 158 nm, since the presence of the surfactant reduces the size of the nanofibers. Wang et al. [89] produced core–shell nanofibers of PVP (shell) and SiO_2 (core) by co-electrospinning, after what the PVP was removed by calcination, leaving highly ordered mesoporous SiO_2 nanofibers with pores of approximately 6.2 nm in diameter and that were successfully tested as catalysts.

In turn, solution blow spinning denotes a technique in which a jet of compressed air is injected coaxially to a polymer jet, being air fed in the outer nozzle inlet, while the polymer solution is fed in the inner nozzle (Fig. 2.4). In this method, the electrostatic force is dismissed, since the compressed air jet provides the force that stretches the polymer jet into the fibers. Several materials have been employed to produce micro- and nanofibers by solution blow spinning, including poly(methyl methacrylate) (PMMA), polystyrene (PS), poly(lactic acid) (PLA) with possible applications in biological tissues [53], soy protein [74], cellulose [115], poly(acrylonitrile) (PAN) aiming at air filtration [114], and polyurethane (PU) with possible applications in filters, protective textiles, scaffolds, wound dressings, and battery components [61].

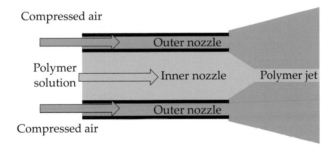

Figure 2.4   Scheme of blow spinning.

2.1.4   Centrifugal Spinning

The centrifugal spinning fabricates micro- and nanofibers of polymers from solutions or melts using centrifugal forces. In addition to the production of fibers by a wide range of polymers, it is possible to use metal, ceramic, or composites [72]. The centrifugal spinning apparatus is flexible and easy to implement, composed of a spinneret, where the needles are fixed, and the collector, as can be seen in Fig. 2.5. In order to produce the fibers, the material is placed in the spinneret, where through the centrifugal force, it flows through the needles. The diameter of the material jet at the needle exit is reduced well below the needle diameter due to the angular velocity. The fibers continue to be elongated during the trajectory of the jet around the spinneret. The greatest reduction in the fiber diameter happens in this period. After that, a collection system is used to gather the fibers [58].

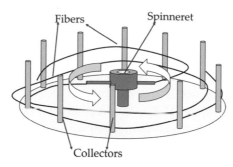

Figure 2.5 Schematic illustration of the centrifugal spinning equipment.

The operational parameters (solution temperature, needle–collector distance, diameter of the needle, solution concentration, angular velocity), material properties (evaporation rate of the solvent, viscosity, surface tension, and elasticity), and equipment components (type of collector and spinneret) influence the production of the fiber [44, 58, 72]. The fibers manufactured by centrifugal spinning highlight for no use of an electric field, large-scale production, promote the use of different materials, and lower the cost of fiber. In addition, fiber manufacturing is not dependent of the solution conductivity and can be performed with high solution concentrations [70].

Some applications of centrifugal spinning include anodes for batteries [2, 18, 104], adsorption [37, 68], air and liquid filtration [54, 70, 71], wound dressing [96], pharmaceutical application [4, 50], colorimetric sensor for food [87], fibrous color dosimeter [35], and tissue engineering [45, 51].

2.1.5   Melt Spinning and Melt Blow Spinning

In melt spinning, granules or pellets of the polymer are extruded. In some cases, the melted polymer passes through a filter [62] or screen [30] before feeding the nozzle or spinneret, from which the nanofibers are produced. In the particular case of melt blow spinning, analogous to the solution blow spinning, a stream of air is fed coaxially to the polymer jet in the nozzle or spinneret, in order to draw the polymer from it without using a system of bobbins [105]. Figure 2.6 exemplifies both methods.

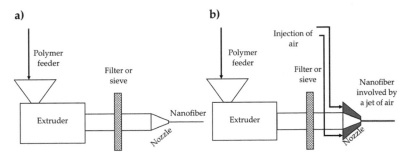

a)

b)

Figure 2.6 Representation of melt spinning (a) and melt blow spinning (b).

Some technologies have been employed to manufacture nanofibers by these techniques, especially to obtain filter media for air and liquid filtration. Hwang et al. [30] produced polypropylene nanofibers for air filtration using a melt spinning apparatus that charged the fibers after passing through the nozzle. The authors tested the collection performance of the spun filter media by comparing them to commercial filter media made of glass fiber according to ISO and European standards (EN 1822-1:2009), reporting that the spun filter media provided higher collection efficiency and less pressure drop than that obtained using the commercial products, before and after dust loading. Yu et al. [106] produced poly(lactic acid) filter media containing tourmaline particles by melt blow spinning, charging the filter media after their manufacturing with a corona charger. Their work reported that the addition of tourmaline increased the mechanical and electrical properties and subsequently the efficiency to collect NaCl nanoparticles (300–500 nm) at air velocity of 5.3 cm/s up to 88.6%. Pu et al. [62] produced polypropylene micro- and nanofibers using an electrically assisted melt blow spinning apparatus, in which the collector and a cooper frame located after the nozzle were connected to a DC power source, reporting that the fibers produced with this method were 40% finer than that produced by the conventional technique (from 1.69 to 0.96 µm). When tested as air filters, they collected up to 98.969% of Di-ethyl-hexyl-sebacat (DEHS) particles with 2.5 µm of diameter at 10.0 m^3/h of air. In addition, Liu et al. [43] injected N_2 in the nozzle of a melt spinning system in order to produce hollow fiber membranes of polyurethane for liquid microfiltration applications, obtaining pore sizes from 180 to 220 nm and providing relative permeability down to 0.49% when tested with 5 wt.‰ of $CaCO_3$ at transmembrane pressure of 0.5 bar and cross-flow velocity of 0.86 m/s. Of particular interest for air filtration is the application of corona discharge in the melt blow spinning of polypropylene fibers containing nanoparticles of polyhedral oligomeric silsesquioxanes (POSS) produced by Song et al. [77]. The manufactured fibers obtained electret properties that provided collection efficiencies up to 97.36% for monodisperse polystyrene particles (0.3 μm) at gas velocity of 32.0 cm/s.

2.1.6   Dry Spinning, Wet Spinning, and Gel Spinning

Other "spinning techniques" are available for specific situations, as in the case when the polymer is not a thermoplastic and/or it is not possible to dissolve it in a volatile solvent, hence melt spinning and conventional electrospinning are not feasible. In dry spinning (Fig. 2.7a), a solution of the non-thermoplastic polymer with sufficiently high concentration passes through the nozzle inserted in a chamber in which a stream of hot gas promotes the solidification of the produced fibers by evaporating the solvent while a system of winding bobbins draws them [105]. Examples of fibers manufactured by dry spinning comprise polyamic acid [97], $CaCl_2$-polyamide 6 [103], and silk protein [78].

Analogously, in wet spinning (Fig. 2.7b), which is feasible for non-thermoplastic polymers that can be dissolved in non-volatile solvents, the chamber in which the polymer jet is drawn is filled with an appropriate non-solvent hence called coagulation bath and that promotes the precipitation and solidification of the fibers [105]. Chitosan–poly(vinyl alcohol) (PVA) for biological applications [88], poly(vinylidene fluoride) (PVDF) [31] aiming at piezoelectric applications, and metal–organic composites with graphene oxide with promising applications for energy storage and electronic devices [109] have been produced by this technique.

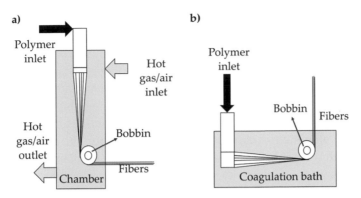

Figure 2.7 Representation of dry and wet spinning.

Systems of nanofiber production combining both methods (dry and wet spinning) have been developed and called gel spinning or dry-jet wet spinning, among similar denominations. In these systems, in general, there is a distance or gap between the nozzle and the coagulation bath, in a way that the polymer jet contacts the air or a gas before entering in contact with the solvent in order to allow the stress relaxation of polymer chains [60]. This technique was employed to produce, for example, lyocell fibers from bagasse with N-methylmorpholine-N-oxide (NMMO) [98], fibers of PVDF doped with naphthopyran-functionalized silica nanoparticles for photochromic applications [60], PVA fibers with cellulose whiskers of nanometric scale for improved mechanical resistance [86], nanoporous membranes of poly(ether sulfone) (PES) with different additives for air filtration of nanoparticles [39]), and multi-channel capillary membranes of poly(piperazine-amide)/PES for low-pressure nanofiltration [6].

2.1.7 Template Synthesis

Template synthesis is a generic term to describe the manufacturing of structures based on a template or mold (Fig. 2.8). For example, Feng et al. [17] developed poly(acrylonitrile) (PAN) nanofibers (104.6 nm) with high hydrophobicity (water contact angles above 170°C) using an anodic aluminum oxide membrane as the template and extruding a PAN solution through the pores of the membrane. Metal nanowires focusing on electrochemical applications were manufactured by Graves et al. [20] with polycarbonate membranes as the templates, in which silver nanoparticles were sputtered before being immersed in an electroless copper bath to form an electrode layer. The authors promoted the electrodeposition of copper and then dissolved the template in dichloromethane, obtaining copper layers between 300 and 500 nm. Hollow carbon nanofibers doped with Cu_2O nanoparticles were developed by Li et al. [40], focusing their application on non-enzymatic glucose biosensors and using anodic aluminum oxide membrane as the template. The experimental procedure yielded nanofibers with 60 nm of diameter, which presented good selectivity to be used as a glucose biosensor. Hollow spheres of δ-MnO_2 for supercapacitors were also manufactured by template synthesis using SiO_2 [95], forming a core–shell structure that yields hollow δ-MnO_2 spheres with diameter of 380 nm and thickness of 40 nm after thermal decomposition of the SiO_2 particles.

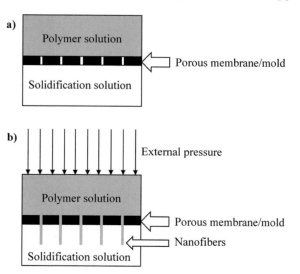

Figure 2.8 Representation of template synthesis before (a) and during (b) the process.

Ceramic nanofibers have been manufactured by a sort of hybrid process combining the electrospinning of a polymeric material (used as the template) from a polymer solution with nanoparticles of the ceramic materials, then removing the template by dissolving them in an appropriate solvent or by calcination in order to obtain porous, ceramic membranes. Table 2.1 exhibits a series of studies using this method with different templates and nanoparticles.

Table 2.1 Template synthesis of ceramic materials after electrospinning

Ceramic Material	Template	Method of Template Removal	Application	Source
V_2O_5, TiO_2, Ta_2O_5, $TaNbO_5$, Nb_2O_5, $VTiO_2$	Pluronic P-123; Brij 76	Calcination	Smart papers and textiles; catalysis	[48]
ZnO–SnO_2	Poly(vinylpyridine) (PVP); Pluronic P-123	Calcination	Ethanol sensing	[76]
Borosilicate glass	Poly(vinyl butyral) (PVB); Pluronic F127	Calcination	Bone tissue regeneration	[19]
SiO_2	Pluronic F127	Calcination	Adsorption-based filtration	[94]
SiO_2	PVP; Pluronic F127	Calcination	Catalysis; drug release	[89]
SiO_2 with $NaYF_4$:Yb^{3+}, Er^{3+} nanocrystals	PVP; Pluronic P-123	Calcination	Drug release on its bioactive, luminescent, and porous properties	[27]
SiO_2	PVP; cetyltrimethyl ammonium bromide (CTAB)	Dissolution in HCl/ acetone	Adsorption-based filtration	[84]
TiO_2	Poly(methylmethacrylate) (PMMA); CTAB	Calcination	Photocatalysis, sensing, photovoltaic cells	[14]
$CaTiO_3$	PVP; Pluronic F127	Calcination	Drug release	[110]
Al/Fe_2O_3	PVP	Calcination	Propellants, explosives, and pyrotechnics	[92]
ZrO_2	PVP; Pluronic F127	Calcination	Thermal insulation; catalysis	[13]

2.1.8 Phase Separation

The manufacturing of nanofibers and nanostructures by phase separation follows a series of steps, which are basically the dissolution of the raw material (essentially a polymer) in a solvent, gelation, extraction of the solvent, cooling, and drying (Fig. 2.9). Ma and Zhang [47] used this technique to build up a 3D nanofibrous network (diameters from 50 to 500 nm) for tissue regeneration applications. The authors prepared solutions of poly(L-lactic acid) (PLLA), poly(D-L-lactic acid-co-glycolic acid) (PLGA), and poly(D-L-lactic acid) (PDLLA) in appropriate organic solvents, promoted gelation by freezing, exchanged the solvent by immersion in distillated water, then froze and froze-dried the gel, obtaining nanostructures with porosities up to 98.5%.

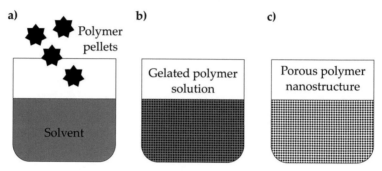

Figure 2.9 Scheme of general phase separation steps: polymer dissolution (a), solution gelation (b), and solvent extraction (c).

Different technologies have been employed to improve this process through the years, whether by using ternary [26] or quaternary systems [28] to modify the physical properties needed for the separation, by adding gelatin particles in the solution to control the size of the pores inside the structures or by drying using supercritical CO_2 [69]. In addition, He et al. [25] proposed a process of nonsolvent-induced phase separation to coat commercial polypropylene separators (for lithium-ion batteries) with poly-p-phenylene terephthamide (PPTA) to be operated in industrial scale, and which was composed by steps of PPTA polymerization, blade casting, coagulation, water cleansing, and drying. The nanocomposites produced by this method exhibited better cycling performances than the original commercial separator. Other nonsolvent-induced phase separation methods can be encountered in Guillen et al. [21] and Hao et al. [22].

Although it has been used to produce nanomaterials for tissue engineering as presented by most of the abovementioned works, phase separation has also been applied to create membranes for liquid filtration [52, 64, 85] and even being associated with electrospinning systems in different ways to create porous, submicrometric fibers or nanofibers focusing on different applications, as drug delivery and catalysis [10, 32, 56, 93].

2.1.9 Self-Assembly

In this technique, nanofibers or nanostructures are constructed by the successive addition of smaller units or molecules that are put together by means of intermolecular forces, in general, as represented in Fig. 2.10. These units could be copolymers of synthetic materials [15, 41, 42, 99], animal proteins, and peptides [12, 23, 111], or even it is possible to use DNA origami to construct metallic nanostructures, such as gold nanorods, [59] and fabricate nanostructures from other nanostructures, as the nanofibers made of ferrocene–diphenylalanine nanospheres produced by Wang et al. [91].

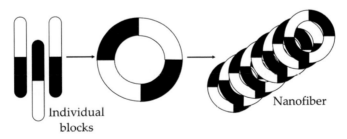

Individual
blocks

Nanofiber

Figure 2.10 Representation of self-assembly.

The methodology of self-assembly nanostructures is diverse, ranging from simple incubation of the precursor solution on a surface and subsequent washing with water [12] or an association between incubation, addition of a solvent, dialysis, and washing [15], shaking and incubation [91], to heating and UV irradiation followed by addition of a solvent and centrifugation [41, 90].

Nanostructures have been manufactured by self-assembly focusing on applications such as liquid filtration. Roebuck and Tremblay [67] studied the effect of the CTAB concentration and the ratio between ethanol and water on the fabrication of boehmite nanostructures in a mixture of water and alcohol, reporting that these parameters affected the porosity, size, and shape of the structures, whose thickness varied between 100 and 200 nm. Karunakaran et al. [34] developed ultrafiltration membranes of poly(styrene-block-polyethylene oxide) (PS-b-PEO) combining self-assembly and water-induced phase separation, yielding membranes with pore sizes from 20 to 30 nm that provided water flux of approximately 800 L/m² h bar.

2.2   Techniques of Characterization

In order to know, manipulate, control, and optimize the operating conditions of the processes that manufacture nanofibers and nanostructures according to the desirable application, it is necessary to characterize not only the morphological properties of the produced materials, but also the properties of the raw materials and their properties in solution or when they are melted. Viscosity, surface tension, and even electrical conductivity are generally measured when working with polymer solutions. Analyses of thermal and mechanical resistance may also be required depending on the applications. For membranes and filter media for air and liquid applications, structural properties of the layers of nanofibers are necessary, such as the size distribution and orientation of the fibers, size distribution of pores, porosity, as well as performance characteristics, such as air or water permeability, collection or retention efficiency, retention time, etc. For catalytic and photo-sensing applications, specific techniques and tests are also needed. Therefore, Table 2.2 summarizes some common techniques used to characterize nanofibers and nanostructures that were performed in some of the works presented over this chapter.

Table 2.2   Techniques to characterize polymer solutions, melted polymers, the structure of the produced nanomaterials and their performance in different applications

Characterization technique	Property evaluated	Source
Transmission electron microscopy (TEM)	Fiber/pore diameter or structural properties	C: [89]; D: [24, 57]/E: [1, 9]/ SA: [34]/T: [20, 95]
Atomic force microscopy (AFM)	Fiber/pore diameter or surface	D: [57]/G: [6]/SA: [34]
Scanning electron microscopy (SEM) or field emission scanning electron microscopy (FESEM)	Fiber/pore diameter and surface	C: [89]; CS: [37, 50, 54]/D: [24]/DS: [78, 103]/E: [1, 8, 9]/G: [6, 98]/MB: [43, 106]/P: [22, 64, 85]/SA: [34, 67]/SB: [70, 114]/T: [20, 95]/W: [31, 88]
Differential scanning calorimetry (DSC)	Melting point	D: [24]/DS: [103]/G: [98]/ MB: [106]/SA: [34]/W: [88]
Thermogravimetric analysis (TGA)	Thermal resistance	CS: [37, 54]/E: [1]/G: [98]
Wide-angle X-ray diffraction (WAXD) or X-ray powder diffraction (XRD)	Crystalline structure	C: [89]; D: [24]/DS: [78, 103]/E: [1]/G: [98]/MB: [106]/W: [31]
Size exclusion chromatography	Molecular weight distribution	D: [24]
Energy dispersive X-ray spectrometry (EDS/EDX)	Elemental composition and distribution over the fibers surface	E: [1, 8, 9]/P: [64]
Fourier-transform infrared spectroscopy (FTIR)	Chemical bonds and functional groups present in the fibers	CS: [37, 50]/DS: [97, 100]/E: [1, 9]/G: [98]/P: [22]/SA: [67]/W: [31, 88]
Cloud point determination	Phase separation conditions	P: [52, 85]
Water contact angle	Hydrophilicity	CS: [50]/P: [64]

(Continued)

Table 2.2   (*Continued*)

Characterization technique	Property evaluated	Source
Conductometry	Electrical conductivity of the solution	E: [8, 9]/SA: [67]
Tensiometry	Surface tension of the solution	E: [8, 9]
Viscosimetry	Viscosity of the solution	E: [8, 9]/SB: [61]
Surface charge density	Charging characteristics of the membrane/filter media	MB: [106]
Porosimetry	Porosity/pore size distribution of the membrane/filter media	E: [8]/MB: [43, 106]/SB: [61, 114]
N_2 adsorption/ desorption tests	Pore size distribution; specific surface area	E: [108]/T: [95]
Antibacterial tests	Bactericidal activity of the membrane/ filter media	E: [1, 8, 9]
Electrochemical tests	Electrochemical performance	E: [108]/T: [95]
Liquid filtration tests	Performance of the filter media in liquid filtration	CS: [54]/G: [6]/MB: [43]/P: [22, 64, 85]/SA: [34]
Air-filtration tests	Performance of the filter media in air filtration	E: [8, 9, 46]/G: [39]/M: [30]/ MB: [77, 106]
Mechanical tests (compression, tensile, shear tests)	Mechanical resistance of fibers layer	CS: [50]/D: [82, 83]/DS: [78, 97]/G: [98]/MB: [106]/P: [64, 85]/W: [31, 88]
Catalytical/ Photocatalytical tests	Performance as catalyst	C: [89]/E: [102]

C: Co-electrospinning; CS: Centrifugal spinning; D: Drawing; DS: Dry spinning; E: Electrospinning; G: Gel spinning; M: Melt spinning; MB: Melt blow spinning; P: Phase separation; SA: Self-assembly; SB: Solution blow spinning; T: Template synthesis; W: Wet spinning

2.3   Conclusion

This chapter aimed to present some of the most recent advances in the science and technology of nanofibers and nanostructures that have been achieved in the scientific field over the globe. Different techniques have been described that are suitable for several applications, according to the properties of the raw materials and the specifications of the final products. Attention was given for liquid and air filtration, bringing some results of collection or retention efficiencies and different configurations and systems built up to optimize the manufacturing of filter media and membranes used for air conditioners and respiratory face masks, for example. However, different applications have been pointed out for each manufacturing technique in order to demonstrate their wide applicability. Finally, some of the most common characterization techniques and performance tests have been presented in order to summarize the usual efforts required to develop and optimize nanomaterials for specific applications, especially nanofibers.

References

1. Abdo, H. S., Khalil, K. A., Al-Deyab, S. S., Altaleb, H., and Sherif, E.-S. M. (2013). Antibacterial effect of carbon nanofibers containing Ag nanoparticles, *Fiber. Polym.*, **14** (12), pp. 1985–1992. Doi: 10.1007/s12221-013-1985-3.

2. Agubra, V. A., De la Garza, D., Gallegos, L., and Alcoutlabi, M. (2016). Force spinning of polyacrylonitrile for mass production of lithium-ion battery separators, *J. Appl. Polym. Sci.*, **133** (1), pp. 1–8. Doi: 10.1002/app.42847.

3. Ahmed, F. E., Lalia, B. S., Hilal, N., and Hashaikeh, R. (2014). Underwater superoleophobic cellulose/electrospun PVDF–HFP membranes for efficient oil/water separation, *Desalination*, **344**, pp. 48–54. Doi: 10.1016/j.desal.2014.03.010.

4. Akia, M., Rodriguez, C., Materon, L., Gilkerson, R., and Lozano, K. (2019). Antibacterial activity of polymeric nanofiber membranes impregnated with Texas sour orange juice, *Eur. Polym. J.*, **115**, pp. 1–5. Doi: 10.1016/j.eurpolymj.2019.03.019.

5. Aytac, Z., Sen, H. S., Durgun, E., and Uyar, T. (2015). Sulfisoxazole/cyclodextrin inclusion complex incorporated in electrospun

hydroxypropyl cellulose nanofibers as drug delivery system, *Colloids Surf. B*, **128**, pp. 331–338. Doi: 10.1016/j.colsurfb.2015.02.019.

6. Back, J. O., Spruck, M., Koch, M., Mayr, L., Penner, S., and Rupprich, M. (2017). Poly(piperazine-amide)/PES composite multi-channel capillary membranes for low-pressure nanofiltration, *Polymers*, **9** (12), 654. Doi: 10.3390/polym9120654.

7. Bortolassi, A. C. C. (2019). Desenvolvimento e avaliação de meios filtrantes com nanofibras eletrofiadas e agentes bactericidas aplicados na filtração de ar (Doctoral Dissertation). Federal University of São Carlos, São Carlos, Brazil. Retrieved on March 27 from https://repositorio.ufscar.br/handle/ufscar/11282.

8. Bortolassi, A. C. C., Guerra, V. G., Aguiar, M. L., Soussan, L., Cornu, D., Miele, P., and Bechelany, M. (2019). Composites based on nanoparticle and pan electrospun nanofiber membranes for air filtration and bacterial removal, *Nanomaterials*, **9** (12), pp. 1740. Doi: 10.3390/nano9121740.

9. Bortolassi, A. C. C., Nagarajan, S., Lima, B. A., Guerra, V. G., Aguiar, M. L., Huon, V., Soussan, L., Cornu, D., Miele, P., and Bechelany, M. (2019). Efficient nanoparticles removal and bactericidal action of electrospun nanofibers membranes for air filtration, *Mater. Sci. Eng. C.*, **102**, pp. 718–729. Doi: 10.1016/j.msec.2019.04.094.

10. Buttaro, L. M, Drufva, E., and Frey, M. W. (2014). Phase separation to create hydrophilic yet non-water soluble PLA/PLA-b-PEG fibers via electrospinning, *J. Appl. Polym. Sci.*, **131** (19), 41030 pp. Doi: 10.1002/app.41030.

11. Cacciotti, I., Fortunati, E., Puglia, D., Kenny, J. M., and Nanni, F. (2014). Effect of silver nanoparticles and cellulose nanocrystals on electrospun poly(lactic) acid mats: Morphology, thermal properties and mechanical behavior, *Carbohydr. Polym.*, **103**, pp. 22–31. Doi: 10.1016/j.carbpol.2013.11.052.

12. Chang, J., Peng, X.-F., Hijji, K., Cappello, J., Ghandehari, H., Solares, S. D., and Seog, J. (2011). Nanomechanical stimulus accelerates and directs the self-assembly of silk-elastin-like nanofibers, *J. Am. Chem. Soc.*, **133**, pp. 1745–1747. Doi: 10.1021/ja110191f.

13. Chattopadhyay, S., Bysakh, S., Saha, J., and De, G. (2018). Electrospun ZrO_2 nanofibers: Precursor controlled mesopore ordering and evolution of garland-like nanocrystal arrays, *Dalton Trans.*, **47**, 5789. Doi: 10.1039/c8dt00415crsc.li/Dalton.

14. Choi, K.-I., Lee, S. H., Park, J.-Y., Choi, D.-Y., Hwang, C.-H., Lee, I.-H., and Chang, M. H. (2013). Fabrication and characterization of hollow TiO_2

fibers by microemulsion electrospinning for photocatalytic reactions, *Mater. Lett.*, **112**, pp. 113–116. Doi: 10.1016/j.matlet.2013.08.101.

15. De Moel, K., Alberda Van Ekenstein, G. O. R., Nijland, H., Polushkin, E., and Ten Brinke, G. (2001). Polymeric nanofibers prepared from self-organized supramolecules, *Chem. Mater.*, **13**, 4580–4583. Doi: 10.1021/cm0110932.

16. Deng, L., Ye, H., Li, X., Li, P., Zhang, J, Wang, X., Zhu, M., and Hsiao, B. S. (2018). Self-roughened omniphobic coatings on nanofibrous membrane for membrane distillation, *Sep. Purif. Technol,* **206**, pp. 14–25. Doi: 10.1016/j.seppur.2018.05.035.

17. Feng, L., Shuhong, L., Huanjun, L., Zhai, J., Song, Y., Jiang, L., and Zhu, D. (2002). Super-hydrophobic surface of aligned polyacrylonitrile nanofibers, *Angew. Chem. Int. Ed.*, **41** (7), pp. 1221–1223. Doi: 10.1002/1521-3773(20020402)41:7<1221::AID-ANIE1221>3.0.CO;2-G.

18. Flores, D., Villarreal, J., Lopez, J., and Alcoutlabi, M. (2018). Production of carbon fibers through Forcespinning® for use as anode materials in sodium ion batteries, *Mat. Sci. Eng. B-Adv.*, **236–237**, pp. 70–75. Doi: 10.1016/j.mseb.2018.11.009.

19. Gao, C., Gao, Q., Bao, X., Li, Y., Teramoto, A., and Abe, K. (2011). Preparation and in vitro bioactivity of novel mesoporous borosilicate bioactive glass nanofibers, *J. Am. Ceram. Soc.*, **94** (9), pp. 2841–2845. Doi. 10.1111/j.1551-2916.2011.04434.x.

20. Graves, J. E., Bowker, M. E. A., Summer, A., Greenwood, A., Ponce De León, C., and Walsh, F. C. (2018). A new procedure for the template synthesis of metal nanowires, *Electrochem. Commun.*, **87**, pp. 58–62. Doi: 10.1016/j.elecom.2017.11.022.

21. Guillen, G. R., Ramon, G. Z., Kavehpour, H. P., Kaner, R. B., and Hoek, E. M. V. (2013). Direct microscopic observation of membrane formation by nonsolvent induced phase separation, *J. Membrane Sci.*, **431**, pp. 212–220. Doi: 10.1016/j.memsci.2012.12.031.

22. Hao, Y., Sano, R., Shimomura, A., Matsuyama, H., and Maruyama, T. (2014). Reorganization of the surface geometry of hollow-fiber membranes using dip-coating and vapor-induced phase separation, *J. Membrane Sci.*, **460**, pp. 229–240. Doi: 10.1016/j.memsci.2014.02.039.

23. Hartgerink, J. D., Beniash, E., and Stupp, S. I. (2001). Self-assembly and mineralization of peptide-amphiphile nanofibers, *Science*, **294** (5547), pp. 1684–1688. Doi: 10.1126/science.1063187.

24. Hasegawa, T. and Mikuni, T. (2014). Higher-order structural analysis of nylon-66 nanofibers prepared by carbon dioxide laser supersonic

drawing and exhibiting near-equilibrium melting temperature, *J. Appl. Polym. Sci.*, **131**, 40361. Doi: 10.1002/APP.40361.

25. He, L., Qiu, T., Xie, C., and Tuo, X. (2018). A phase separation method toward PPTA–polypropylene nanocomposite separator for safe lithium ion batteries, *J. Appl. Polym. Sci.*, **134**, 46697. Doi: 10.1002/APP.46697.

26. He, L., Zhang, Y., Zeng, X., Qan, D., Liao, S., Zeng, Y., Lu, J., and Ramakrishna, S. (2009). Fabrication and characterization of poly(L-lactic acid) 3D nanofibrous scaffolds with controlled architecture by liquid–liquid phase separation from a ternary polymer–solvent system, *Polymer*, **50**, pp. 4128–4138. Doi: 10.1016/j.polymer.2009.06.025.

27. Hou, Z., Li, C., Ma, P., Li, G., Cheng, Z., Peng, C., Yang, D., Yang, P., and Lin, J. (2011). Electrospinning preparation and drug-delivery properties of an up-conversion luminescent porous NaYF$_4$:Yb^{3+},Er^{3+}@Silica fiber nanocomposite, *Adv. Funct. Mater.*, **21**, pp. 2356–2365. Doi: 10.1002/adfm.201100193.

28. Hsu, S.-H., Huang, S., Wang, Y.-C., and Kuo, Y.-C. (2013). Novel nanostructured biodegradable polymer matrices fabricated by phase separation techniques for tissue regeneration, *Acta Biomater.*, **9**, pp. 6915–6927. Doi: 10.1016/j.actbio.2013.02.012.

29. Hussain, C. M. (2020) *Handbook of Nanomaterials for Manufacturing Applications* (Elsevier, New Jersey).

30. Hwang, S., Roh, J., and Park, W. M. (2018). Comparison of the relative performance efficiencies of melt-blown and glass fiber filter media for managing fine particles, *Aerosol Sci. Technol.*, **52** (4), pp. 451–458. Doi: 10.1080/02786826.2017.1423274.

31. Jeong, K., Kim, D. H., Chung, Y. S., Hwang, S. K., Hwang, H. Y., and Kim, S. S. (2018). Effect of processing parameters of the continuous wet spinning system on the crystal phase of PVDF fibers, *J. Appl. Polym. Sci.*, **135**, 45712. Doi: 10.1002/APP.45712.

32. Jiang, Y., Fang, D., Song, G., Nie, J., Chen, B., and Ma, G. (2013). Fabrication of core–shell nanofibers by single capillary electrospinning combined with vapor induced phase separation, *New J. Chem.*, **37**, 2917. Doi: 10.1039/c3nj00654a.

33. Johnson, D. F., Druce, J. D., Birch C., and Grayson, M. L. (2009). A quantitative assessment of the efficacy of surgical and N95 masks to filter influenza virus in patients with acute influenza infection, *Clin. Infect. Dis.*, **49** (2), pp. 275–277. Doi: 10.1086/600041.

34. Karunakaran, M., Nunes, S. P., Qiu, X., and Yu, H. (2014). Isoporous PS-b-PEO ultrafiltration membranes via self-assembly and water-induced

phase separation, *J. Membrane Sci.*, **453**, pp. 471–477. Doi: 10.1016/j. memsci.2013.11.015.

35. Kinashi, K., Iwata, T., Tsuchida, H., Sakai, W., and Tsutsumi, N. (2018). Composite resin dosimeters: A new concept and design for a fibrous color dosimeter, *ACS Appl. Mater. Interfaces*, **10** (14), pp. 11926–11932. Doi: 10.1021/acsami.8b00251.

36. Koyama, H., Watanabe, Y., and Suzuki, A. (2014). Poly(p-phenylene sulfide) nanofibers prepared by CO_2 laser supersonic drawing, *J. Appl. Polym. Sci.*, **131**, 40922. Doi: 10.1002/APP.40922.

37. Kummer, G., Schonhart, C., Fernandes, M. G., Dotto, G. L., Missio, A. L., Bertuol, D. A., and Tanabe, E. H. (2018). Development of nanofibers composed of chitosan/Nylon 6 and tannin/Nylon 6 for effective adsorption of Cr(VI), *J. Polym. Enrivon.*, **26**, pp. 4073–4084. Doi: 10.1007/s10924-018-1281-9.

38. Li, L., Shang, L., Li, Y., and Yang, C. (2014). Three-layer composite filter media containing electrospun polyimide nanofibers for the removal of fine particles, *Fiber. Polym.*, **18** (4), pp. 749–757. Doi: 10.1007/s12221-017-1094-9.

39. Li, M., Feng, Y., Wang, K., Yong, W. F., Yu, L., and Chung, T.-S. (2017). Novel hollow fiber air filters for the removal of ultrafine particles in PM2.5 with repetitive usage capability, *Environ. Sci. Technol.*, **51**, pp. 10041–10049. Doi: 10.1021/acs.est.7b01494.

40. Li, X., Wang, C., Huang, X., Zhang, T., Wang, X., Min, M., Wang, L., Huang, H., and Hsiao, B. S. (2018). Anionic surfactant-triggered Steiner geometrical poly(vinylidene fluoride) nanofiber/nanonet air filter for efficient particulate matter removal, *ACS Appl. Mater. Interfaces*, **10** (49), pp. 42891–42904. Doi: 10.1021/acsami.8b16564.

41. Liu, G., Ding, J., Qiao, L., Guo, A., Dymov, B. P., Gleeson, J. T., Hashimoto, T., and Saijo, K. (1999). Polystyrene-block-poly(2-cinnamoylethyl methacrylate) nanofibers: Preparation, characterization, and liquid crystalline properties, *Chem. Eur. J.*, **5** (9). Doi: 10.1002/(SICI)1521-3765(19990903)5:9<2740::AID-CHEM2740>3.0.CO;2-V.

42. Liu, G., Qiao, L., and Guo, A. (1996). Diblock copolymer nanofibers, *Macromolecules*, **29**, pp. 5508–5510. Doi: 10.1021/ma9604653.

43. Liu, M., Xiao C., and Hu, X. (2012). Fouling characteristics of polyurethane-based hollow fiber membrane in microfiltration process, *Desalination*, **298**, pp. 59–66. Doi: 10.1016/j.desal.2012.05.002.

44. Lu, Y., Li, Y., Zhang, S., Xu, G., Fu, K., Lee, H., and Zhang, X. (2013). Parameter study and characterization for polyacrylonitrile nanofibers

fabricated via centrifugal spinning process, *Eur. Polym. J.*, **49** (12), pp. 3834–3845. Doi: 10.1016/j.eurpolymj.2013.09.017.

45. Lukášová, V., Buzgo, M., Vocetková, K., Sovková, V., Doupník, M., Himawan, E., Staffa, A., Sedláček, R., Chlup, H., Rustichelli, F., Amler, E., and Rampichová, M. (2019). Needleless electrospun and centrifugal spun poly-ε-caprolactone scaffolds as a carrier for platelets in tissue engineering applications: A comparative study with hMSCs, *Mater. Sci. Eng. C*, **97**, pp. 567–575. Doi: 10.1016/j.msec.2018.12.069.

46. Lv, D., Wang, R., Tang, G., Mou, Z., Lei, J., Han, J., de Smedt, S., Xiong, R., and Huang, C. (2019). Eco-friendly electrospun membranes loaded with visible-light response nano-particles for multifunctional usages: High-efficient air filtration, dye scavenger and bactericide, *ACS Appl. Mater. Interfaces*, **11** (13), pp. 12880–12889. Doi: 10.1021/acsami.9b01508.

47. Ma, P. X. and Zhang, R. (1999). Synthetic nano-scale fibrous extracellular matrix, *Biomed. Mater. Res.*, **46**, pp. 60–72. Doi: 10.1002/(SICI)1097-4636(199907)46%3A1<60%3A%3AAID-JBM7>3.0.CO%3B2-H.

48. Macías, M., Chacko, A., Ferraris, J. P., and Balkus Jr., K. J. (2005). Electrospun mesoporous metal oxide fibers, *Microporous Mesoporous Mater.*, **86**, pp. 1–13. Doi: 10.1016/j.micromeso.2005.05.053.

49. Majidi, S. S., Slemming-Adamsen, P., Hanif, M., Zhang, Z., Wang, Z., and Chen, M. (2018). Wet electrospun alginate/gelatin hydrogel nanofibers for 3D cell culture, *Int. J. Biol. Macromol.*, **118**, pp. 1648–1654. Doi: 10.1016/j.ijbiomac.2018.07.005.

50. Mamidi, N., Romo, I. L., Barrera, E. V., and Elías-Zúñiga, A. (2018). High throughput fabrication of curcumin embedded gelatin-polylactic acid forcespun fiber-aligned scaffolds for the controlled release of curcumin, *MRS Commun.*, **8** (4), pp. 1395–1403. Doi: 10.1557/mrc.2018.193.

51. Mamidi, N., Romo, I. L., Gutiérrez, H. M. L., Barrera, E. V., and Elías-Zúñiga, A. (2018). Development of forcespun fiber-aligned scaffolds from gelatin–zein composites for potential use in tissue engineering and drug release, *MRS Commun.*, **8** (3), pp. 885–892. Doi: 10.1557/mrc.2018.89.

52. M'barki, O., Hanafia, A., Bouyer, D., Faur, C., Sescousse, R., Delabre, U., Blot, C., Guenoun, P., Deratani, A., Quemener, D., and Pochat-Bohatier, C. (2014). Greener method to prepare porous polymer membranes by combining thermally induced phase separation and crosslinking of poly(vinyl alcohol) in water, *J. Membrane Sci.*, **458**, pp. 225–235. Doi: 10.1016/j.memsci.2013.12.013.

53. Medeiros, E. S., Glenn, G. M., Klamczynski, A. P., Orts, W. J., and Mattoso, L. H.C. (2009). Solution blow spinning: A new method to produce

micro- and nanofibers from polymer solutions, *J. Appl. Polym. Sci.*, **113**, pp. 2322–233. Doi: 10.1002/app.30275.

54. Missau, J., Rocha, J. G., Dotto, G. L., Bertuol, D. A., Ceron, L. P., and Tanabe, E. H. (2018). Purification of crude wax using a filter medium modified with a nanofiber coating, *Chem. Eng. Res. Des.*, **136**, pp. 734–743. Doi: 10.1016/j.cherd.2018.06.031.

55. Nain, A. S., Amon, C., and Sitti, M. (2006). Proximal probes based nanorobotic drawing of polymer micro/nanofibers, *IEEE Trans. Nanotechnol.*, **5** (5), pp. 499–510. Doi: 10.1109/TNANO.2006.880453.

56. Nayani, K., Katepalli, H., Sharma, C. S., Sharma, A., Patil, S., and Venkataraghavan, R. (2012). Electrospinning combined with nonsolvent-induced phase separation to fabricate highly porous and hollow submicrometer polymer fibers, *Ind. Eng. Chem. Res.*, **51**, pp. 1761–1766. Doi: 10.1021/ie2009229.

57. Ondarçuhu, T. and Joachim, C. (1998). Drawing a single nanofibre over hundreds of Microns, *Europhys. Lett.*, **42** (2), pp. 215–220. Doi: 10.1209/epl/i1998-00233-9.

58. Padron, S., Fuentes, A., Caruntu, D., and Lozano, K. (2013). Experimental study of nanofiber production through forcespinning, *J. Appl. Phys.*, **113**, 024318. Doi: 10.1063/1.4769886.

59. Pal, S., Deng, Z., Wang, H., Zou, S., and Liu, Y. (2011). DNA directed self-assembly of anisotropic plasmonic nanostructures, *J. Am. Chem. Soc.*, **133**, pp. 17606–17609. Doi: 10.1021/ja207898r.

60. Pinto, T. V., Cardoso, N., Costa, P., Sousa, C. M., Durães, N., Silva, C., Coelho, P. J., Pereira, C., and Freire, C. (2019). Light driven PVDF fibers based on photochromic nanosilica@naphthopyran fabricated by wet spinning, *Appl. Surf. Sci.*, **470**, pp. 951–958. Doi: 10.1016/j.apsusc.2018.11.203.

61. Polat, Y., Pampal, E. S., Stojanovska, E., Simsek, R., Hassanin, A., Kilic, A., Demir, A., and Yilmaz, S. (2016). Solution blowing of thermoplastic polyurethane nanofibers: A facile method to produce flexible porous materials, *J. Appl. Polym. Sci.*, **13**, 43025. Doi: 10.1002/APP.43025.

62. Pu, Y., Zheng, J., Chen, F., Long, Y., Wu, H., Li, Q., Yu, S., Wang, X., and Ning, X. (2018). Preparation of polypropylene micro and nanofibers by electrostatic-assisted melt blown and their application, *Polymers*, **10**, 959. Doi: 10.3390/polym10090959.

63. Rahimi, M. and Mokhtari, J. (2018). Modeling and optimization of waterproof-breathable thermo-regulating core-shell nanofiber/net structured membrane for protective clothing applications, *Polym. Eng. Sci.*, **58**, 10. Doi: 10.1002/pen.24776.

64. Rajabzadeh, S., Liang, C., Ohmukai, Y., Maruyama, T., and Matsuyama, H. (2012). Effect of additives on the morphology and properties of poly(vinylidene fluoride) blend hollow fiber membrane prepared by the thermally induced phase separation method, *J. Membrane Sci.*, **423–424**, pp. 189–194. Doi: 10.1016/j.memsci.2012.08.013.

65. Ramakrishna, S., Fujihara, K., Teo, W.-E., Lim, T.-C., and Ma, Z. (2005) *An Introduction to Electrospinning of Nanofibers* (World Scientific, Singapore).

66. Reneker, D. H. and Chun, I. (1996). Nanometre diameter fibres of polymer, produced by electrospinning, *Nanotechnology*, **7** (3), pp. 216–223. Doi: 10.1088/0957-4484/7/3/009.

67. Roebuck, K. and Tremblay, A. Y. (2016). The self-assembly of twinned boehmite nanosheets into porous 3D structures in ethanol–water mixtures, *Colloids Surf. A*, **495**, pp. 238–247. Doi: 1 0.1016/j. colsurfa.2016.01.025.

68. Rostamian, R., Firouzzare, M., and Irandoust, M. (2019). Preparation and neutralization of forcespun chitosan nanofibers from shrimp shell waste and study on its uranium adsorption in aqueous media, *React. Funct. Polym.*, **143**, 104335. Doi: 10.1016/j. reactfunctpolym.2019.104335.

69. Salerno, A., Fernández-Gutiérrez, M., Del Barrio, J. S. R., and Domingo, C. (2015). Bio-safe fabrication of PLA scaffolds for bone tissue engineering by combining phase separation, porogen leaching and scCO$_2$ drying, *J. Supercrit. Fluids*, **97**, pp. 238–246. Doi: 10.1016/j. supflu.2014.10.029.

70. Salussoglia, A. I. P., Tanabe, E. H., and Aguiar, M. L. (2019). Characterization and micro and nanoparticles filtration evaluation of Pan nanofibers produced by centrifugal spinning, *Enciclopédia Biosfera*, **16** (30), pp. 103–112. Doi: 10.18677/EnciBio_2019B10.

71. Salussoglia, A. I. P., Tanabe, E. H., and Aguiar, M. L. (2020). Evaluation of a vacuum collection system in the preparation of PAN fibers by forcespinning for application in ultrafine particle filtration, *J. Appl. Polym. Sci.*, pp. 49334. Doi: 10.1002/app.49334.

72. Sarkar, K., Gomez, C., Zambrano, S., Ramirez, M., Hoyos, E., Vasquez, H., and Lozano, K. (2010). Electrospinning to Forcespinning™, *Mat. Today*, **13** (11), pp. 12–14. Doi: 10.1016/S1369-7021(10)70199-1.

73. Shalaby, T., Hamad, H., Ibrahim, E., Mahmoud, O., and Al-Oufy, A. (2018). Electrospun nanofibers hybrid composites membranes for highly efficient antibacterial activity, *Ecotox. Environ. Safe.*, **162** (30), pp. 354–364. Doi: 10.1016/j.ecoenv.2018.07.016.

74. Sinha-Ray, S., Zhang, Y., Yarin, A. L., Davis, S. C., and Pourdeyhimi, B. (2011). Solution blowing of soy protein fibers, *Biomacromolecules*, **12**, pp. 2357–2363. Doi: 10.1021/bm200438v.

75. Son, W. K., Youk, J. H., and Park, W. H. (2006). Antimicrobial cellulose acetate nanofibers containing silver nanoparticles, *Carbohydr. Polym.*, **65** (4), pp. 430–434. Doi: 10.1016/j.carbpol.2006.01.037.

76. Song, X., Wang, Z., Liu, Y., Wang, C., and Li, L. (2009). A highly sensitive ethanol sensor based on mesoporous $ZnO-SnO_2$ nanofibers, *Nanotechnol.*, **20**, 075501. Doi: 10.1088/0957-4484/20/7/075501.

77. Song, X., Zhou, S., Wang, Y., Kang, W., and Cheng, B. (2012). Mechanical and electret properties of polypropylene unwoven fabrics reinforced with POSS for electret filter materials, *J. Polym. Res.*, **19**, 9812. Doi: 10.1007/s10965-011-9812-2.

78. Sun, M., Zhang, Y., Zhao, Y., Shao, H., and Hu, X. (2012). The structure–property relationships of artificial silk fabricated by dry-spinning process, *J. Mater. Chem.*, **22**, 18372. Doi: 10.1039/c2jm32576d.

79. Sung, A. D., Sung, J. A. M., Thomas, S., Hyslop T., Gasparetto, C., Long, G., Rizzieri, D., Sullivan, K. M., Corbert, K., Broadwater, G., Chao, N. J., Horwitz, M. E. (2016). Universal mask usage for reduction of respiratory viral infections after stem cell transplant: A prospective trial, *Clin. Infect. Dis.*, **63** (8), pp. 999–1006. Doi: 10.1093/cid/ciw451.

80. Suzuki, A. and Hayashi, H. (2013). Ethylene tetrafluoroethylene nanofibers prepared by CO_2 laser supersonic drawing, *EXPRESS Polym. Lett.*, **7** (6), pp. 519–527. Doi: 10.3144/expresspolymlett.2013.48.

81. Suzuki, A., Mikuni, T., and Hasegawa, T. (2014). Nylon 66 nanofibers prepared by CO_2 laser supersonic drawing, *J. Appl. Polym. Sci.*, **131**, 40015. Doi: 10.1002/APP.40015.

82. Suzuki, A. and Ohta, K. (2018). Mechanical properties of poly(ethylene terephthalate) nanofiber three-dimensional structure prepared by CO_2 laser supersonic drawing, *J. Appl. Polym. Sci.*, **135**, 4. Doi: 10.1002/APP.45763.

83. Suzuki, A. and Tanizawa, K. (2009). Poly(ethylene terephthalate) nanofibers prepared by CO_2 laser supersonic drawing, *Polymer*, **50**, pp. 913–921. Doi: 10.1016/j.polymer.2008.12.037.

84. Taha, A. A., Qiao, J., Li, F., and Zhang, B. (2012). Preparation and application of amino functionalized mesoporous nanofiber membrane via electrospinning for adsorption of Cr^{3+} from aqueous solution, *J. Environ. Sci.*, **24** (4), pp. 610–616. Doi: 10.1016/S1001-0742(11)60806-1.

85. Tanaka, T., Nishimoto, T., Tsukamoto, K., Yoshida, M., Kouya, T., Taniguchi, M., and Lloyd, D. R. (2012). Formation of depth filter microfiltration membranes of poly(l-lactic acid) via phase separation, *J. Membrane Sci.*, **396**, pp. 101–109. Doi: 10.1016/j.memsci.2012.01.002.

86. Uddin, A. J., Araki, J., Gotoh, Y., and Takatera, M. (2011). A novel approach to reduce fibrillation of PVA fibres using cellulose whiskers, *Text. Res. J.*, **81** (5), pp. 44–458. Doi: 10.1177/0040517511399967.

87. Valdez, M., Gupta, S. K., Lozano, K., and Mao, Y. (2019). ForceSpun polydiacetylene nanofibers as colorimetric sensor for food spoilage detection, *Sens. Actuators B Chem.*, **297**, 126734. Doi: 10.1016/j.snb.2019.126734.

88. Vega-Cázarez, C. A., López-Cervantes, J., Sánchez-Machado, D. I., Madera-Santana, T. J., Soto-Cota, A., and Ramírez-Wong, B. (2018). Preparation and properties of chitosan–PVA fibers produced by wet spinning, *J. Polym. Environ.*, **26**, pp. 946–958. Doi: 10.1007/s10924-017-1003-8.

89. Wang, H., Wu, D., Li, D., Niu, Z., Chen, Y., Tang, D., Wu, M., Cao, J., and Huang, Y. (2011). Fabrication of continuous highly ordered mesoporous silica nanofibre with core/sheath structure and its application as catalyst carrier, *Nanoscale*, **3**, 3601. Doi: 10.1039/c1nr10547g.

90. Wang, J.-Y., Wang, Y., Sheiko, S. S. Betts, D. E., and Desimone, J. M. (2012). Tuning multiphase amphiphilic rods to direct self-assembly, *J. Am. Chem. Soc.*, **134**, pp. 5801–5806. Doi: 10.1021/ja2066187.

91. Wang, Y., Huang, R., Qi, W., Wu, Z., Su, R., and He, Z. (2013). Kinetically controlled self-assembly of redox-active ferrocene–diphenylalanine: From nanospheres to nanofibers, *Nanotechnology*, **24**, 465603. Doi: 10.1088/0957-4484/24/46/465603.

92. Wang, Z., Zhang, T.-F., Ge, Z., and Luo, Y.-J. (2015). Morphology-controlled synthesis of Al/Fe$_2$O$_3$ nano-composites via electrospinning, *Chin. Chem. Lett.*, **26**, pp. 1535–1537. Doi: 10.1016/j.cclet.2015.07.017.

93. Wei, Z., Zhang, Q., Wang, L., Wang, X., Long, S., and Yang, J. (2013). Porous electrospun ultrafine fibers via a liquid–liquid phase separation method, *Colloid. Polym. Sci.*, **291**, pp. 1293–1296. Doi: 10.1007/s00396-012-2858-9.

94. Wu, Y.-N., Li, F., Wu, Y., Jia, W., Hannam, P., Qiao, J., and Li, G. (2011). Formation of silica nanofibers with hierarchical structure via electrospinning, *Colloid Polym. Sci.*, **289**, pp. 1253–1260. Doi: 10.1007/s00396-011-2455-3.

95. Xiao, W., Zhou, W., Yu, H., Pu, Y., Zhang, Y., and Hu, C. (2018). Template synthesis of hierarchical mesoporous d-MnO$_2$ hollow

microspheres as electrode material for high-performance symmetric supercapacitor, *Electrochimica Acta*, **264**, pp. 1–11. Doi: 10.1016/j.electacta.2018.01.070.

96. Xu. F., Weng, B., Materon, L. A., Gilkerson, R., and Lozano, K. (2014). Large-scale production of a ternary composite nanofiber membrane for wound dressing applications, *J. Bioact. Compat. Pol.*, **29** (6), pp. 646–660. Doi: 10.1177/0883911514556959.

97. Xu, Y., Wang, S., Li, Z., Xu, Q., and Zhang, Q. (2013). Polyimide fibers prepared by dry-spinning process: Imidization degree and mechanical properties, *J. Mater. Sci.*, **48**, pp. 7863–7868. Doi: 10.1007/s10853-013-7310-0.

98. Yamamoto, A., Uddin, A. J., Gotoh, Y., Nagura, M., and Iwata, M. (2011). Dry jet-wet spinning of bagasse/N-methylmorpholine-N-oxide hydrate solution and physical properties of lyocell fibres, *J. Appl. Polym. Sci.*, **119**, pp. 3152–3161. Doi: 10.1002/app.33151.

99. Yan, X., Liu, G., Liu, F., Tang, B. Z., Peng, H., Pakhomov, A. B., and Wong, C. Y. (2001). Superparamagnetic triblock copolymer/Fe_2O_3 hybrid nanofibers, *Angew. Chem. Int.*, **40** (19), pp. 3593–3596. Doi: 10.1002/1521-3773(20011001)40:19<3593::AID-ANIE3593>3.0.CO;2-U.

100. Yang, S. and Lee, G. W. (2005). Electrostatic enhancement of collection efficiency of the fibrous filter pretreated with ionic surfactants., *J. Air Waste Manage.*, **55** (5), pp. 594–603. Doi: 10.1080/10473289.2005.10464655.

101. Yang, S. and Lee, G. W. M. (2005). Filtration characteristics of a fibrous filter pretreated with anionic surfactants for monodisperse solid aerosols, *J. Aerosol Sci.*, **36** (4), pp. 419–437. Doi: 10.1016/j.jaerosci.2004.10.002.

102. Yang, Y., Wen, J., Wei, J., Xiong, R., Shi, J., and Pan, C. (2013). Polypyrrole-decorated Ag-TiO_2 nanofibers exhibiting enhanced photocatalytic activity under visible-light illumination, *ACS Appl. Mater. Interfaces,* **5** (13), pp. 6201–6207. Doi: 10.1021/am401167y.

103. Yang, Z., Yin, H., Li, X., Liu, Z., and Jia, Q. (2010). Study on dry spinning and structure of low mole ratio complex of calcium chloride-polyamide 6, *J. Appl. Polym. Sci.*, **118**, pp. 1996–2004. Doi: 10.1002/app.32176.

104. Yanilmaz, M. and Zhang, X. (2015). Polymethylmethacrylate/polyacrylonitrile membranes via centrifugal spinning as separator in Li-ion batteries, *Polymers*, **7** (4), pp. 629–643. Doi: 10.3390/polym7040629.

105. Yarin, A. L., Pourdeyhimi, B., and Ramakrishna, S. (2014) *Fundamentals and Applications of Micro- and Nanofibers*. (Cambridge University Press, U.K.).

106. Yu, B., Han, J., Sun, H., Zhu, F., Zhang, Q., and Kong, J. (2015). The preparation and property of poly(lactic acid)/tourmaline blends and melt-blown nonwoven, *Polym. Compos.*, **36** (2), pp. 264–271. Doi: 10.1002/pc.22939.

107. Yu, D.-G., Williams, G. R., Gao, L.-D., Annie Bligh, S. W., Yang, J.-H., and Wang, X. (2012). Coaxial electrospinning with sodium dodecylbenzene sulfonate solution for high quality polyacrylonitrile nanofibers, *Colloids Surf. A*, **396**, pp. 161–168. Doi: 10.1016/j.colsurfa.2011.12.063.

108. Yu, S. and Myung, N. V. (2018). Minimizing the diameter of electrospun polyacrylonitrile (PAN) nanofibers by design of experiments for electrochemical application, *Electroanalysis*, **30** (10), pp. 2330–2338. Doi: 10.1002/elan.201800368.

109. Zhang, L., Liu, W., Shi, W., Xu, X., Mao, J., Li, P., Ye, C., Yin, R., Ye, S., Liu, X., Cao, X., and Gao, C. (2018). Boosting lithium storage properties of MOF derivatives through a wet-spinning assembled fiber strategy, *Chem. Eur. J.*, **24**, pp. 13792–13799. Doi: 10.1002/chem.201802826.

110. Zhang, Q., Li, X., Ren, Z., Han, G., and Mao, C. (2015). Synthesis of CaTiO$_3$ nanofibers with controllable drug-release kinetics, *Eur. J. Inorg. Chem.*, **27**, pp. 4532–4538. Doi: 10.1002/ejic.201500737.

111. Zheng, Z., Chen, P., Xie, M., Wu, C., Luo, Y., Wang, W., and Jiang, J. (2016). Cell environment-differentiated self-assembly of nanofibers, *J. Am. Chem. Soc.*, **138**, pp. 11128–11131. Doi: 10.1021/jacs.6b06903.

112. Zhou, F. L., Li, Z., Gough, J. E., Hubbard Cristinacce, P. L., and Parker, G. J. M. (2018). Axon mimicking hydrophilic hollow polycaprolactone microfibres for diffusion magnetic resonance imaging, *Mater. Design*, **137**, pp. 394–403. Doi: 10.1016/j.matdes.2017.10.047.

113. Zhou, S. S., Lukula, S., Chiossone, C., Nims, R. W., Suchmann, D. B., and Ijaz, M. K. (2018). Assessment of a respiratory face mask for capturing air pollutants and pathogens including human influenza and rhinoviruses, *J. Thorac. Dis.*, **10** (93), pp. 2059–2069. Doi: 10.21037/jtd.2018.03.10.

114. Zhuang, X., Jia, K., Cheng, B., Guan, K., Kang, W., and Ren, Y. (2013). Preparation of polyacrylonitrile nanofibers by solution blowing process, *J. Eng. Fiber. Fabr.*, **8** (1), pp. 88–93. Doi: 10.1177/155892501300800111.

115. Zhuang, X., Yang, X., Shi, L., Cheng, B., Guan, K., and Kang, W. (2012). Solution blowing of submicron-scale cellulose fibers, *Carbohydr. Polym.*, **90**, pp. 982–987. Doi: /10.1016/j.carbpol.2012.06.031.

Chapter 3

Risks and Effects on Human Health of Nanomaterials

Gustavo Marques da Costa,[a] Aline Belem Machado,[b]
Daniela Montanari Migliavacca Osório,[b]
Daiane Bolzan Berlese,[b] and
Chaudhery Mustansar Hussain[c]

[a]Instituto Federal de Educação Ciência e Tecnologia Farroupilha (IFFar)
Campus Santo Augusto, Santo Augusto, CEP 98590-000, RS, Brazil
[b]Feevale University, ERS-239, 2755, Novo Hamburgo, CEP 93525-075, RS, Brazil
[c]Department of Chemistry and Environmental Science, New Jersey Institute of
Technology, Newark, NJ, United States
markesdakosta@hotmail.com

Research involving nanotoxicology has shown that interactions of nanomaterials with cells, animals, humans, and the environment are extremely complex. The morphological and physical–chemical properties of nanomaterials have a great impact on the interaction of nanomaterials with cells, in a biological environment, and thus impact their toxicity. Studies suggest that nanomaterials, due to their small size, may have a greater permeability through the skin, mucous membranes, and cell membranes, and may have their toxic

Environmental, Ethical, and Economical Issues of Nanotechnology
Edited by Chaudhery Mustansar Hussain and Gustavo Marques da Costa
Copyright © 2022 Jenny Stanford Publishing Pte. Ltd.
ISBN 978-981-4877-76-3 (Hardcover), 978-1-003-26185-8 (eBook)
www.jennystanford.com

effect magnified, since they have a higher reactivity, mainly due to the increase in the surface area. This chapter aims to present concepts, case studies, and research related to the destination, toxicity, and the effects of nanomaterials on the environment and human health.

3.1 Introduction

3.1.1 Nanomaterials and Their Effects

A nanomaterial, according to the European Community, is "a natural, incidental or manufactured material containing particles, in a non-aggregated or aggregated state, in which 50% or more of the particles in the numerical size distribution, one or more external dimensions are in the size range 1 nm – 100 nm." It is important to take into account that nanomaterials may become emerging environmental contaminants in the not-too-distant future, since there is a growing interest in these types of materials. Since biomolecules (proteins, carbohydrates, lipids, and nucleic acids) and the basic unit of life (cells) have the same size scale as nanomaterials, they can interact when placed in contact with the formation of a bio–nano interface. Depending on the nature of the bio–nano interactions, significant reflections on the higher levels of the organization of the biosystems may occur, due to the interconnectivity between all levels. When the nanoparticles reach the cells, they can be absorbed through the endocytosis, which consists of invagination of the cell wall over the particle until it encompasses it completely. The concern with the inhalation of nanoparticles is mainly due to the fact that the smaller the particle, the more easily it overcomes the natural barriers of the respiratory system, being deposited and accumulated in the alveoli, responsible for the gas exchange of O_2 and CO_2 with the bloodstream.

3.2 Toxicity, Transport, and Destination

Nanomaterials are being discovered every day, and because they make more efficient products feasible, many of these are in the commercialization phase [1]. However, the same properties that make nanomaterials attractive may also be responsible for harmful effects on living organisms. Nanomaterials can have natural origin, for example nanoparticles originating from volcanic emissions that travel great distances through the air and some viruses, or have anthropogenic origin, which are those nanoparticles produced as a result of human activities, such as in refining, smoking, and engines [2]. The European Union's Scientific Committee on Consumer Products has classified the use of nanoparticles into two groups: labile and insoluble. Labels are easily destroyed by predictable physicochemical conditions, in the case of liposomes, lipid nanoparticles, and biodegradable nanoparticles, while insoluble particles, such as carbon nanotubes and metallic nanoparticles, are unable to break down in biological environments [3].

Nanotechnologies can be understood as the human capacity to understand, modify, and control matter on a nanometer scale (10^{-9} m) since biomolecules (proteins, carbohydrates, lipids, and nucleic acids) and the basic unit of life (cells) are included in the same size scale as nanomaterials. In this sense, we can understand that nanomaterials will have impacts on technologies and life sciences, since nanosystems are being planned and built with the ability to interact with the lower levels of biological organization, such as DNA and cells. Therefore, the main motivations for studying the interaction of nanomaterials with biosystems are (1) equivalent nanometric size with the size of the biomacromolecules and cells; (2) capacity for interaction and cellular penetration; (3) chemical surface can be modified (functionalization) [4]; (4) potential for modulation and control of specific biological functions; (5) assessment of toxicity and its impacts on human and environmental health.

Toxicology has great importance in contemporary society, being an area of knowledge essential for sustainable development. Thus, nanotoxicology emerges as a new division within the toxicological sciences, with nanomaterials as the object of study [5, 6]. Toxicology can be defined as the science that studies the adverse effects of agents of a physical, chemical, or biological nature on biosystems, with the goal of treatment, diagnosis, and, mainly, the prevention of intoxication. Also, the concern with nanotoxicity arises as diverse nanomaterials are synthesized, manipulated, and discarded in different environments, whether natural, urban, or industrial, without proper control and regulation.

In the Department of Environmental Health (DSA), one of the aspects of research in nanomaterials is the search for new strategies for assessing toxicity based on predictive and high-performance technologies to assess the potential risk of these materials. In this sense, one of the international collaboration projects in the field of nano-security, currently underway, addresses the occupational risks associated with the production of nanomaterials in the ceramic industry. This project will evaluate the exposure of workers in different scenarios and will characterize nanoparticles produced intentionally and unintentionally during industrial processes. The laboratory will evaluate the biological/toxicological effects of nanoceramic materials transported by air in relevant models in vitro and in vivo [7].

Nanoparticles, after absorption and once inside the organism, as a result of their small dimensions, can travel to the circulatory and lymphatic system, reaching different organs and tissues, including the brain. The deposition of nanomaterials in tissues can trigger an inflammatory response in which cells such as macrophages and neutrophils are recruited to the site of contact, leading to oxidative stress at the cellular level (production of ROS), which, in turn, can cause changes in the genome adjacent cells, producing secondary genotoxic effects. Therefore, for the assessment and risk management of any substance or agent, it is necessary to determine precisely and unambiguously its toxicity, that is, the degree or intensity of the adverse/toxic effect on a given bioindicator/biomarker of exposure.

A study on mice examined the interaction of metallic nanoparticles with biological systems [8]. After oral administration of formulations of different sizes, it was possible to verify an increase in inflammatory responses, as well as toxicity in important organs, liver, and kidneys. In this sense, the need to understand the mechanisms that determine the behavior of nanoparticles is evident, not only for the development of this technology but also to predict toxicological responses to nanomaterials [9].

Another necessary point to be considered in relation to nanomaterials, before their commercialization, is to carry out an assessment of their life cycle—including manufacturing, transportation, use, recycling, and disposal of waste in order to observe its effects on the environment, health, and safety. In addition, due to mobility and persistence in soil, water, and air, bioaccumulation and interactions with chemical and biological materials, products with nanomaterials represent an unprecedented class of manufactured contaminants. However, it is also important to mention that the presence of nanomaterials in the environment does not mean that there will always be the manifestation and observation of adverse or harmful (toxic) effects associated with them. However, it is important to remember that the expression of these effects depends on the characteristics of the exposure and its behavior in the environment. Therefore, the release into the environment of nanoparticles and carbon nanotubes must always be avoided and companies and research laboratories must treat these materials as hazardous, seeking to reduce their use and eliminate them from waste systems.

3.3   Risk of Nanoparticles to Humans

Nanotechnology is a multidisciplinary scientific field based on the development, characterization, production, and application of structures, devices, and systems with shape and size in nanometric scale, and may present chemical, physical–chemical, and behavioral properties different than those presented on larger scales [10, 11].

The human population is exposed to exogenous particles of dimensions in the order of nanometers. Nanomaterials are found in the air that we breathe and in the consumer products we daily use [12].

The exposure to nanoproducts can occur during the various stages of the life cycle in the elaboration/use process, denoting their impact potential from synthesis and production to consumer exposure to these products, with the elimination and consequent accumulation in the environment, thus constituting a source of human exposure through environmental exposure [13].

There is uncertainty about the risk of nanoparticles to humans, which may be more toxic when compared to the same material in a non-nano scale [14, 15].

There are many scientific uncertainties about the understanding of the risks related to nanoparticles and whether they could cause disturbances at molecular and cellular levels. Because of their similarities in size to biological macromolecules, such as proteins, DNA, and phospholipids, nanoparticles can cause inflammation, destruction of brain cells, and injuries.

Nanomaterials in tissues can trigger an inflammation response in which cells such as macrophages and neutrophils are recruited to the contact site, leading to oxidative stress that can cause alterations in the genome of adjacent cells, producing secondary genotoxic effects. And these genetic changes may facilitate the process of transforming cells [16].

In addition to these oxidative DNA damage, the direct genotoxic effects of nanomaterials can also play a decisive role in generating genetic instability and consequent cancerous processes [17].

Another type of adverse effect on the exposure to nanomaterials includes endocrine disruption. These chemical substances can interfere in the hormonal system, altering the natural way of communication of the endocrine system [2].

The biological system has chemical defense mechanisms and biological barriers; however, the nanoparticles have small dimensions, with the ability to move to the circulatory and lymphatic systems, reaching several organs and tissues, including the brain. The inflammatory response that results in the production of reactive oxygen species (ROS) correlates exposure to nanomaterials with the occurrence of tissue damage and, by causing damage to cell genome, has the potential to contribute to the development of neoplasms [18].

The chemical properties of nanomaterials enable their absorption, being metabolized and eliminated or accumulated in the body. These properties can, however, be modified dynamically when under different biological or environmental conditions.

The studies related to the interaction of nanoparticles, doses, and biological systems are of fundamental importance, mainly in the mediators of oxidative stress in the face of cellular injuries, toxicity, and mutagenicity, in order to contribute to the safe use of these materials [19].

When nanoparticles reach the cells, they can be absorbed by the cellular membrane by several processes, one of them being endocytosis, which involves the invagination of the cellular wall over the particles until it completely encompasses them [20]. The transdermal route is also an important entry route, when it comes to the use of cosmetics or personal hygiene products that contain nanomaterials in their composition [2].

The skin has a complex structure, as its function is to be the boundary between the organism and the environment, thus presenting low permeability. Its composition in functional layers makes it efficient in all its actions, from a barrier against pathogens to a water and thermal regulator for the organism.

In the pharmaceutical area, the advances in nanotechnology have enabled the production of drugs with less active principle and more directed to the target of action, as it is possible to involve the drug and release it gradually. These systems have been developed for countless therapeutic applications [21].

After the absorption, and once inside the organism, due to their small size, the nanoparticles have the ability to translocate to the circulatory and lymphatic system, reaching several tissues and organs [22].

The respiratory tract can be divided into three regions: nasopharynx, tracheobronchial, and alveolar [23]. And from there, certain size particles can be deposited in each region; for example, 90% of nanoparticles with 1 nm in diameter are deposited in the nasopharynx region, while only 10% of these nanoparticles are deposited in the tracheobronchial region and almost none reach the alveolar region [24].

The presence of these nanoparticles in the air is due to the process of erosion of manufactured materials or to the production, use, and/or manipulation of nanoparticles in industrial processes. Thus, the inhalation route is the most important human exposure route, mainly in the occupational context [2].

After being deposited in the pulmonary epithelium, the nanoparticles appear to translocate to extrapulmonary sites, reaching other organs by different mechanisms and routes, leading the particles into the bloodstream, or being transported by lymphocytes, resulting in the distribution of nanoparticles throughout the body [5].

The nanoparticles have the ability to overcome biological barriers due to their nanometric size. The nanoparticles can deposit or cross the alveolar barriers and cause great activation of the immune system, fibrosis, or other systematic responses, and it is possible to observe chronic effects, which concerns regarding acute or chronic exposures.

The diseases that have been associated to inhaled nanoparticles are asthma, bronchitis, emphysema, lung cancer, and other neurodegenerative diseases, such as Parkinson's and Alzheimer's diseases. In addition to the passage through the airway, which is more evident, studies confirm that the absorption of inhaled nanoparticles from the olfactory mucosa by the olfactory nerves in the olfactory bulb [22]. What can be concluded is a defense response by inhaling nanoparticles, as they are highly penetrable.

The studies of nanotoxicity related to inhalation point to the adoption of limiting factors in the elaboration, production, and even the correct disposal of nanoparticles, mainly avoiding the emission of these types of material in the atmosphere.

The nanomaterials incorporated in food, dietary supplements, or even food packaging, as well as those from contaminated soil or water, may be absorbed through the intestines of mammals and thus may eventually lead to systemic effects [2]. These effects may arise after oral exposure, as the nanoparticles are distributed to the kidneys, lever, spleen, lungs, brain, and gastrointestinal tract [23].

Upon reaching the blood, nanoparticles can be eliminated by mechanisms other than those that are dependent on the absorption route and in their surface properties. For example, the absorption through the lungs, skin, or intestinal tract can be in the form of particles contained within phagocytic cells, such as macrophages, in individual particles or aggregates of free particles or associated with whey proteins [25].

The energy metabolism can also be a target of toxicity related to nanoparticles. The main concern related to the interaction of nanoparticles is the fact that some nanomaterials are transported across cell membrane, especially in mitochondria [26].

However, whether the nanoparticles enter directly in the mitochondria or are internalized because of the oxidative damage remains unclear. The loss of mitochondrial potential, which is induced by nanoparticles, has important biological effects, especially in decreasing the production of ATP and in the beginning of apoptosis [27, 28].

There is no doubt that nanotechnology offers the prospect of major advances that impact the health and quality of life of human beings. However, the introduction of new materials and chemical substances brings with it some health risks and these issues need to be guided in order to assess the risk/benefit ratio.

3.4 Conclusion

Nanotechnology is part of development, providing a basis for innovation in terms of applications and products, with emphasis on the area of medicine and even consumer products, especially in industrialized countries. Numerous benefits involving nanotechnology are described for improving the environment and human health. However, these same characteristics that make nanoparticles important for the development of technology can become undesirable when released into the environment. Manufactured nanomaterials contribute to induce oxidative stress states that are associated with aging and the appearance of various diseases, inflammatory processes, destruction of brain cells, and precancerous lesions.

In this sense, not all manufactured nanomaterials are equal in terms of their toxic potential. There are several in vitro and in vivo studies that indicate adverse biological effects of nanomaterials with a potential impact on human health. Due to the complexity associated with their physical and chemical properties, the characterization of the genotoxicity of these materials and the comparison of the results of several studies is a challenge for scientists and legislators. Therefore, it is recommended to apply the precautionary principle, as an instrument capable of enabling the management of risks produced by nanotechnologies and carrying out research involving the use and disposal of nanomaterials.

Websites

1. http://nanodb.dk/en/analysis/consumer-products/#chartHashsection

2. *nanodb.dk/en*

References

1. Hussain, C. M. (2020). *The ESLI Handbook of Nanotechnology: Risk Safety, ESLI and Commercialization*, 1st Ed. (Wiley).

2. Louro, H., Borges, T., and Silva, M. J. (2013). Nanomateriais manufaturados: Novos desafios para a saúde pública. *Revista Portuguesa de Saúde Pública*, **31**(2), pp. 188–200.

3. Rosen, J. E., Chan, L., Shieh, D.-B., and Gu, F. X. (2011). Iron oxide nanoparticles for targeted cancer imaging and diagnostics. *Nanomedicine*, **84**, pp. 41–45.

4. Hussain, C. M. (2020). *Handbook of Functionalized Nanomaterials for Industrial Applications*, 1st Ed. (Elsevier).

5. Oberdorster, G., Oberdorster, E., and Oberdorster, J. (2005) Nanotoxicology: An emerging discipline evolving from studies of ultrafine particles. *Environ Health Perspect*, **113**(7), pp. 823–839.

6. Nel, A., Xia, T., Mädler, L., and Li, N. (2006). Toxic potential of materials at the nanolevel. *Science*, **311**(5761), pp. 622–627.

7. SNS - Serviço Nacional de Saúde. Available from: <http://www.insa.min-saude.pt/avaliacao-da-toxicidade-dos-nanomateriais/> (accessed June 2020).

8. Park, M. V., Neigh, A. M., Vermeulen, J. P., Liset, J. J. F., Henny, W. V., Jacob, J. B., Henk, L., and Wim, H. J. (2011). The effect of particle size on the cytotoxicity, inflammation, developmental toxicity and genotoxicity of silver nanoparticles. *Biomaterials*, **32**(36), pp. 9810–9817.

9. Fadeel, B. and Garcia-Bennett, A. (2010). Better safe than sorry: Understanding the toxicological properties of inorganic nanoparticles manufactured for biomedical applications. *Adv Drug Deliv Rev*, **62**, pp. 362–374.

10. Meyer, M. and Persson, O. (1998). Nanotechnology-interdisciplinarity, patterns of collaboration and differences in application. *Scientometrics*, **42**(2), pp. 195–205.

11. Allarakhia, M. and Walsh, S. (2012). Analyzing and organizing nanotechnology development: Application of the institutional analysis

development framework to nanotechnology consortia. *Technovation*, **32**(3–4), pp. 216–226.

12. da Silva, L. H., Viana, A. R., Baldissera, M. D., Nascimento, K., Sagrillo, M. R., and Luchese, C. (2014). Revisão bibliográfica sobre relações entre nanomateriais, toxicidade e avaliação de riscos: A emergência da nanotoxicologia. *Disciplinarum Scientia/ Saúde*, **15**(1), pp. 19–30.

13. Ferreira, A. P. and da Silva Sant'Anna, L. (2015). A Nanotecnologia e a questão da sua regulação no Brasil: Impactos à saúde e ao ambiente. *Revista Uniandrade*, **16**(3), pp. 119–128.

14. McComas, K. A. and Besley, J. C. (2011). Fairness and nanotechnology concern. *Risk Anal: An Int. J.*, **31**(11), pp. 1749–1761.

15. Kuempel, E. D., Geraci, C. L., and Schulte, P. A. (2012). Risk assessment and risk management of nanomaterials in the workplace: Translating research to practice. *Ann. Occup. Hyg.*, **56**(5), pp. 491–505.

16. Singh, N., Manshian, B., Jenkins, G. J., Griffiths, S. M., Williams, P. M., Maffeis, T. G., Wright, C. J., Doak, S. H., and Wright, C. (2009). NanoGenotoxicology: The DNA damaging potential of engineered nanomaterials. *Biomaterials*, **30**(23–24), pp. 3891–3914.

17. IARC – International Agency for Research on Cancer. (2010). *Carbon Black, Titanium Dioxide and Talc*. Lyon: World Health Organization.

18. Stone, V. and Donaldson, K. (2006). Signs of stress. *Nat. Nanotechnol.*, **1**(1), pp. 23.

19. Elmore, S. (2007). Apoptosis: A review of programmed cell death. *Toxicol. Pathol.*, **35**(4), pp. 495–516.

20. Auffan, M., Decome, L., Rose, J., Orsiere, T., De Meo, M., Briois, V., Chaneac, C., Olivi, L., Berge-Lefranc, J-L., Botta, A., Wiesner, M. R., and Bottero, J.-Y. (2006). In vitro interactions between DMSA-coated maghemite nanoparticles and human fibroblasts: A physicochemical and cyto-genotoxical study. Environ. Sci. Technol., **40**(14), pp. 4367–4373.

21. Schaffazick, S. R., Guterres, S. S., Freitas, L. D. L., and Pohlmann, A. R. (2003). Caracterização e estabilidade físico-química de sistemas poliméricos nanoparticulados para administração de fármacos. *Química Nova*, **26**(5), pp. 726–737.

22. Buzea, C., Pacheco, I. I., and Robbie, K. (2007). Nanomaterials and nanoparticles: Sources and toxicity. *Biointerphases*, **2**(4), pp. MR17–MR71.

23. Forbe, T., Gárcia, M., and Gonzalez, E. (2011) Potential risks of nanoparticles Riscos potenciais do nanopartículas, *Ciência e Tecnologia de Alimentos*, **31**, pp. 835–842.

24. Moghimi, S. M., Hunter, A. C., and Murray, J. C. (2005). Nanomedicine: Current status and future prospects. *The FASEB J.*, **19**(3), pp. 311–330.

25. Borm, P. J., Robbins, D., Haubold, S., Kuhlbusch, T., Fissan, H., Donaldson, K., Schins, R., Stone, V., Kreyling, W., Lademann, J., Krutmann, J., Warheit, D., and Oberdorster, E. (2006). The potential risks of nanomaterials: A review carried out for ECETOC. *Part. Fibre Toxicol.*, **3**(1), 11.

26. Foley, S., Crowley, C., Smaihi, M., Bonfils, C., Erlanger, B. F., Seta, P., and Larroque, C. (2002). Cellular localisation of a water-soluble fullerene derivative. *Biochem. Biophys. Res. Commun.* **294**(1), pp. 116–119.

27. Hiura, T. S., Li, N., Kaplan, R., Horwitz, M., Seagrave, J. C., and Nel, A. E. (2000). The role of a mitochondrial pathway in the induction of apoptosis by chemicals extracted from diesel exhaust particles. *J. Immunol.*, **165**(5), pp. 2703–2711.

28. Teodoro, J. S., Simões, A. M., Duarte, F. V., Rolo, A. P., Murdoch, R. C., Hussain, S. M., and Palmeira, C. M. (2011). Assessment of the toxicity of silver nanoparticles in vitro: A mitochondrial perspective. *Toxicol. In Vitro*, **25**(3), pp. 664–670.

Chapter 4

Bioavailability and Toxicity of Manufactured Nanoparticles in Terrestrial Environments

**Aline Belém Machado,[a] Gustavo Marques da Costa,[b]
Daniela Montanari Migliavacca Osório,[a]
Daiane Bolzan Berlese,[a] and
Chaudhery Mustansar Hussain[c]**

[a]*Feevale University, ERS-239, 2755, Novo Hamburgo,
CEP 93525-075, RS, Brazil*
[b]*Instituto Federal de Educação Ciência e Tecnologia Farroupilha (IFFar)
Campus Santo Augusto, Santo Augusto, CEP 98590-000, RS, Brazil*
[c]*Department of Chemistry and Environmental Science,
New Jersey Institute of Technology, Newark, NJ, United States*
markesdakosta@hotmail.com

The contribution of nanoparticles in the environment can be caused by natural and/or anthropogenic sources. The so-called anthropogenic sources are mainly related to industrial processes or activities that involve burning fossil fuels and generate large amounts of particulate matter. Therefore, it is important to discern between two types of nanoparticles from anthropogenic sources: engineered

Environmental, Ethical, and Economical Issues of Nanotechnology
Edited by Chaudhery Mustansar Hussain and Gustavo Marques da Costa
Copyright © 2022 Jenny Stanford Publishing Pte. Ltd.
ISBN 978-981-4877-76-3 (Hardcover), 978-1-003-26185-8 (eBook)
www.jennystanford.com

nanoparticles, manufactured for incorporation in materials (nanocomposites and nanomaterials in general), and non-engineered nanoparticles, mainly from burning fossil fuels. Considering the range of products based on engineered nanomaterials available on the market, these can now be considered the main source of input of these materials in the environment and, for this reason, due attention must be given to their manufacturing process, transport, storage, and disposal. Also, environmental properties such as water chemistry (pH, ionic strength, redox potential, hardness, and organic content), soil, and sediment type can alter the bioavailability of nanoparticles. Among the main nanomaterials are organic ones such as carbon nanotubes, metallic ones (mainly metal oxides), and quantum dots, such as biological nanomarkers [1]. It is also important to mention that the different properties of nanoparticles (high surface-area-to-volume ratio and the small size) in relation to their corresponding bulk material can cause toxicity in living organisms. Concentrations of manufactured nanoparticles have rarely been measured in the environment to date. Various techniques are available to characterize nanoparticles for exposure and dosimetry, although each of these methods has advantages and disadvantages for the ecotoxicologist. The aims of this chapter are to present case studies, research, and advances in bioavailability and toxicity manufactured nanoparticles in terrestrial environments.

4.1 Origin of Manufactured Nanoparticles in Terrestrial Environments

The increasing utilization of manufactured nanoparticles (MNPs) in daily products has led to an uncontrollable release of these materials to the environment [2]. The different life cycle stages of these products, such as usage, recycling, and disposal of product containing nanomaterials, can provide different sources of nanotechnology that can be disposed into the environment, which can occur intentionally (e.g., through the applications of nanomaterials in daily care products) or unintentionally (e.g., by the abrasion of clothing-containing nanomaterials) [3].

The use of products containing MNPs has resulted in their disposal in the environment, and consequently reaching the soil, which is a major sink of these nanoparticles [4, 5]. Also, other sources of MNPs in soil are the usage of these materials in water purification systems [6] and in the application of sewage sludge from wastewater treatment plants in agricultural lands [5], including nano-fertilizers, seed treatment preparations, pesticides, etc. [6]. The application of sewage sludge in agricultural fields has made the agricultural soils a great repository and source of MNPs to water systems through erosion and runoff [7]. Also, in an unintentional manner, lands can be exposed using treated wastewater for irrigation purposes [8].

Nanoparticles can be present in the environment by two sources, natural and anthropogenic. Anthropogenic sources include industrial (due to nanotechnology activities, including accidental release and production) residues from the usage of items with nanomaterials, characterized as intentional release [9, 10], combustion processes, and deposition from atmosphere. Natural sources can include nanominerals, volcanic eruption, which can deposit in soils, and biological processes, among others [9].

The remediation for contaminated soils is an important source of MNPs. The nano-zero-valent iron (nZVI) is one of the most recognized nanoparticles used with this aim. The application of MNPs has been recognized as effective in reducing the contamination by organic and inorganic pollutants [4] (Fig. 4.1).

The complexity of soil matrix limits the characterization and the analysis of nanoparticles [11]. Therefore, modeling becomes an important alternative to estimate the concentration of nanomaterials in terrestrial environments.

Thus, a study performed by Giese et al. [12] evaluated the minimum and maximum release of SiO_2-nano, CeO_2-nano, and nanosilver (Ag-nano) through the application of a mathematical modeling, which included the annual nanomaterial mass contribution to domestic production or the importation of pure nanomaterials or products containing nanomaterials.

The annual estimated production volume of CeO_2-nano in Germany was approximately 22,000 ton/a, while the estimated volume for Ag-nano was below 1000 ton/a. The results of the predicted environmental concentration (PEC) for those three nanomaterials for 2017 were expressed as minimum and maximum, taking into consideration an extreme scenario, assuming that all the possible applications of nanomaterials were simultaneously running in the maximum level, with consequent use and release. They

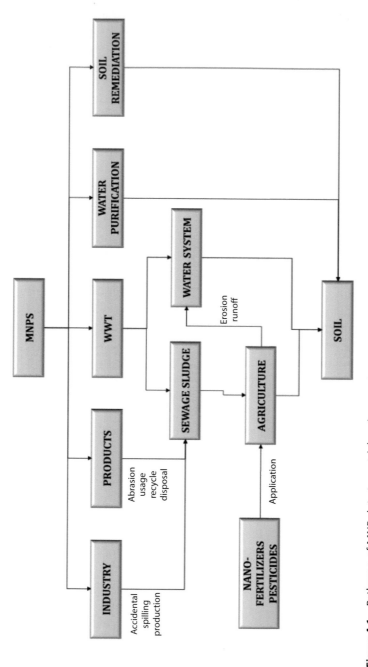

Figure 4.1 Pathways of MNPs into terrestrial environment.
MNPs – manufactured nanoparticles; WWT – wastewater treatment.

also considered two scenarios, one with 100% degradation of the nanomaterial after 1 year, and the other with a 100% persistence of the nanomaterials. Table 4.1 provides the PEC of the studied MNPs, concerning their concentration in agricultural soils and sludge-treated soils.

Table 4.1 PEC for 2017

Nanomaterial	Unit	Minimum	Maximum
SiO_2			
Agricultural soils*	µg/kg	0.00	258.17
Agricultural soils**	µg/kg	0.08	6,981.61
Sludge-treated soil*	µg/kg	0.11	11,603.86
Sludge-treated soil**	µg/kg	2.85	294,408.31
CeO_2			
Agricultural soils*	ng/kg	0.10	2,943.37
Agricultural soils**	ng/kg	2.10	62,109.92
Sludge-treated soil*	ng/kg	0.97	35,069.36
Sludge-treated soil**	ng/kg	25.96	933,959.89
Ag			
Agricultural soils*	ng/kg	0.03	80.30
Agricultural soils**	ng/kg	0.50	1,522.66
Sludge-treated soil*	ng/kg	1.13	2,294.05
Sludge-treated soil**	ng/kg	22.66	45,859.26

Note: Adapted from Ref. [12]. *100% degradation; **100% persistence.

4.2 Bioavailability of Nanomaterials in Terrestrial Environments

In the recent years, the use of nanoparticles has been increasing in many areas, including household and cosmetic products, such as detergents, sunscreens, water filters, and also the industry, as nanocatalysts, solar cells, and electronic devices [13–15]. Another segment where these nanoparticles are used is agrobusiness, with the use of the so-called "precision agriculture," which can aggregate and adapt advanced technologies for a better production yield [16].

Although after the use of these nanoparticles, their behavior in the environment is not known for sure, mainly in water and soil, trying to understand what the mechanisms that these nanoparticles will establish with these environmental matrices is still a challenge. An important feature in this process is to know how these nanoparticles will behave in different environmental conditions, because it is known that in the environment there are many complexities that, until today, we are trying to understand their mechanisms and processes. However, how to understand these interactions is a challenge to the scientific community nowadays.

We have the terrestrial environment, which we can simplify as the soil, an extremely complex compartment with different characteristics, which receives impacts from different natures, giving the use of land for different applications, such as housing, food supply, disposal of waste, and even the survival of species. Thus, what will be the behavior of these nanoparticles in this compartment? What are the processes and interactions we have? What will be the impact of these nanoparticles? Anyway, these are the questions that still require answers, but somehow, we will try to explain a little.

As the soil is a complex compartment, with solid complex mixtures from millimeters to the size of nanometric particles, so many factors of the soil can influence these processes, such as pH, organic matter, ionic strength, in addition to biological factors such as microbial activity.

In this complex matrix, which is the soil, we have to think: How will the process of bioavailability of nanoparticles in the soil be? Which processes are important and, in some way, can make soil nanoparticles available to other compartments of the environment, such as water and plants? Thus, chemical transformations such as oxidation, reduction, and dissolution act in an important way to evaluate the bioavailability processes of nanoparticles in the environment.

Metal oxide nanoparticles (MeO-NPs) are made up of metal oxides in the nanoscale range, which includes synthetized particles and from occurrence of anthropogenic or natural processes. Their frequent use is related to their crystalline structures and the nature of bonding between the metal and oxygen, which can be ionic, covalent, or metallic, which guarantees them electronic and magnetic properties [17] and also their use in engineering such as fuel cells, plastics, and consumer products [18]. Studies on the fate of MeO-NPs in terrestrial systems are less frequent than in aquatic systems; this may be related to underdeveloped analytical techniques to characterize, track, and execute studies on the transformation processes, based on soil suspensions and not the soil [19].

In the environment, the MeO-NPs are subjected to several processes that can affect their physical–chemical properties, as possible sources, routes of introduction, and transfer mechanisms and transformations to MeO-NPs in the environment [18].

The processes involved in the bioavailability of MeO-NPs in the environment are physical, chemical, and biological, in which we can mention the processes of aggregation, agglomeration, adsorption, deposition, dissolution, sulfide, mineralization, and redox, which are the main processes that affect the fate of these materials in the environment. These mechanisms depend heavily on chemical processes, which can vary in different environmental matrices, water, air, and soil [18]. Therefore, in this chapter, some bioavailability processes in the terrestrial environment will be presented.

And yet the bioavailability of MeO-NPs depends on their stability in the environment [20, 21], and the stable NPs are less bioavailable and also can exhibit less toxic effect for cellular organisms [20]. In the process of synthesis and stabilization of NPs, coating materials are used, which also affect the bioavailability [22–24]. Also the interactions of NPs with coexisting substances, such as other NPs, humic materials, proteins, polysaccharides, and different types of ions, can also affect the adsorption property, and consequently their bioavailability [25, 26].

Another important factor is the interaction of NPs with pollutants, such as herbicides, pesticides, anionic species, and metallic elements present in the soil, which can favor their bioavailability [10], so the effects of coexisting pollutants must also be considered in states of mobility of MeO-NPs in the environment. For example, Al_2O_3-NPs and TiO_2-NPs can diffuse through basic rocks and be transported, so their mobility depends on their behavior, such as surface charge, size, stability, and roughness of the components of the medium [27].

4.2.1 Aggregation/Agglomeration

The nanoparticle aggregation process occurs when there is a cluster of NPs, which can occur through the combination of van der Waals forces and electrostatic repulsion, according to the Derjaguine–Landaue–Verweye–Overbeek (DLVO) theory [28], and other interactions such as hydration forces, magnetic and hydrophobic interactions. In the aggregation process, we have the aggregation of equal NPs, called homoaggregation, and the aggregation of different NPs, called hetroaggregation [25, 29, 30]. In the environment, heteroaggregation is more likely to occur than homoaggregation [31]. In the terrestrial environment, the aggregation of MeO-NPs is associated with interactions that occur between soil particles; thus heteroaggregation is more frequent, due to the complexity of materials that make up the soil [32]. Aggregation can affect the bioavailability of NPs in terrestrial systems, which hinders the diffusion process. For example, the use of ZnO-NPs in soils has been reported due to their heteroaggregation with granules [33–35], which favors their mobility due to the colloidal presence between MeO-NPs and soil particles [35].

The aggregation/agglomeration process in terrestrial environment can originate in the aquatic environment; once the aggregation process in the aquatic environment increases the aggregate size, it favors the deposition process that can occur in the sediment [18, 30].

The soil composition directly influences the aggregation process [36, 37]. A reduced stability of CeO_2-NPs in soils has been reported with high clay content, which can be attributed to the adsorption of NPs on the surface of colloidal materials. On the other hand, the presence of Al_2O_3-NPs promoted an aggregation mainly in acid pH due to their low charge and greater hydrophobic potential [38].

4.2.2 Dissolution and Redox

In soil, we have a slower process of residence elements; among them, we can mention metals. Metallic ions linked to MeO-NPs behave differently in soils, because their behavior is directly related to soil properties and also the age of the compartment [39].

In soil, the processes that act more significantly are dissolution and redox reactions, as these establish physical transfers through the process of soil formation, such as aggregation, diffusion, and morphology that must be identified to evaluate the fate of these MeO-NPs and consequently their bioavailability to the terrestrial environment [18].

The dissolution process that MeO-NPs undergo in the soil is directly related to the increase in rainfall, which provided the soluble fraction of the soil, also called soil solution, which increases the entry of water into the soil pores. However, the kinetic process of the MeO-NPs dissolution in soil depends on this soil's characteristics and also on the way in which the MeO-NPs enter, whether in soluble or powdered form. The study performed by Wang et al. [40] with ZnO-NPs observed a rapid dissolution process in the soil, not being detected after 1 h of incorporation in the soil. So, soil properties, as well as pH, redox potential, ionic strength, diffusion, and dissolution, are important characteristics that interfere in dissolution. Made et al. [18] identified that the dissolution of CeO_2-NPs coated with citrate (size 8 nm) was only significant for pH less than 4. Also, another very relevant aspect is the presence of roots in terrestrial environment, which can potentiate the dissolution process. Dimkpa [41] identified that the dissolution of CuO-NPs (<3.0 to 1 mg/kg) and ZnO-NPs (0.6 to between 1 and 2.2 mg/kg) was more pronounced in the presence of roots of *Triticum aestivum*. But we can also find MeO-NPs, such as TiO_2-NPs, which have low solubility in soils [42].

Natural organic matter (NOM) is also related in the dissolution process of MeO-NPs in the environment [43–48]. Due to its complexity and also the presence of metal ions free from NPs, the dissolution mechanism can be identified. Jiang et al. [44] identified that the dissolution of ZnO-NPs increased with the concentration of NOM (0 and 40 mg/L), and the presence of aromatic carbon in the NOM was a major factor for increasing the dissolution. On the other hand, Odzak et al. [47] reported that the high dissolution of ZnO- NPs and CuO-NPs in freshwater conditions presented low concentration of dissolved organic matter (DOM). Thus, the dissolution of MeO-NPs caused by NOM is related to the concentration range of the NOM in the system and to pH [49].

4.2.3 Adsorption

Adsorption is another process related to MeO-NPs and is directly related to the presence of organic substances in the soil, which absorb MeO-NPs via exchange of binders and hydrophilic processes assisting in the stability of NPs [19]. As the adsorption process is very complex, studies are being performed in order to evaluate these interactions through models between DOM and the NPs. These models can be designed in three stages: (1) external film diffusion of DOM from bulk solution to the surface of engineered NPs, (2) binding of DOM onto the surface of engineered NPs, and (3) internal particle diffusion [50–52]. Therefore, the kinetic and thermodynamic models attempt to describe the interaction between DOM-NPs. And it is also believed that the interactions between DOM-NPs depend on the nature and properties of NPs (coating, size, and shape), DOM (such as which functional groups are present and their structure), chemical characteristics (pH, ionic strength, and inorganic ions coexistence), and environmental conditions (temperature, radiation, etc.) [51, 52]. However, more studies are necessary to comprehend the NP–DOM interactions, which are highly complex and their diverse impacts on environmental processes and biological effects.

4.2.4 Chemical Processes

Chemical processes such as the generation of reactive oxygen species (ROS), surface oxidation, and degradation of material coating can make MeO-NPs available for the terrestrial system [18].

4.3 Toxicity of Manufactured Nanoparticles

A concern regarding the toxicity of nanoparticles lies mainly in the fact that they have never been produced and used in commercial products on such a large scale as they are today. In this way, the risk of reaching different environmental compartments and becoming available is high. Recent studies on the toxicity of nanomaterials were carried out in the last decade of the 20th century, investigating materials that, on a micrometric scale, did not present toxicity, and on a nanometric scale, such as nanoparticles, presented some effect in Mexico.

Studies suggest that nanomaterials, due to their small size, may have a greater permeability through the skin, mucous membranes, and cell membranes, and may have their toxic effect magnified, since they have a higher reactivity, mainly due to the increase in the surface area.

Regarding the study of the toxicity of nanomaterials, there are several ways in which the particles can appear after contact with the environment or living organisms and may be present in their free form as well as in clusters. Toxicity tests can be carried out on cell cultures (in vitro) or with living organisms (in vivo) such as fish, mice, and even humans. Studies have already been carried out on reproductive toxicity and the development of manufactured nanomaterials [53–55].

Therefore, nanotoxicology is an essential tool that must be explored according to the progress of studies and innovative applications of nanometric particles, providing the composition of measures for biological protection to prevent damage to health, mainly related to nanopharmacology, avoiding toxicity.

4.4 Conclusion

Each nanoparticle is unique, depending on its preparation route, size, state of aggregation, stability in a biological medium, chemical nature of the coating, and surface charge. Studies show that around 243 products containing nanotechnology are available on the market, but not all mention that they are based on this technology.

The toxicity of nanoparticles can be affected by the size of nanoparticles, frequency and time of exposure, the vulnerability of the exposed organism, the routes of introduction, and mainly the interaction with the biosystems. The smaller nanoparticles tend to have a greater ability to penetrate cells and, consequently, have toxic effects. In addition, a point to be highlighted is related to the cost of carrying out in vivo studies, which often need to be considered in dollars. Therefore, it is necessary to carry out studies with longer exposure and evaluation periods in physiological systems not yet investigated.

Websites

1. http://www. raeng.org.uk/publications/reports/nanoscience-and nanotechnologies-opportunities
2. http://www.nanoreg.eu/

References

1. Hussain, C. M. (2020). *Handbook of Functionalized Nanomaterials for Industrial Applications*, 1st Ed. (Elsevier).
2. Hussain, C. M. (2020). The ESLI Handbook of Nanotechnology: Risk Safety, ESLI and Commercialization, 1st Ed. (Wiley).
3. Gottschalk, F. and Nowack, B. (2011). The release of engineered nanomaterials to the environment. *J. Environ. Monit.*, **13**(5), pp. 1145–1155.
4. Pan, B. and Xing, B. (2012). Applications and implications of manufactured nanoparticles in soils: A review. *Eur. J. Soil Sci.*, **63**(4), pp. 437–456.
5. Torrent, L., Marguí, E., Queralt, I., Hidalgo, M., and Iglesias, M. (2019). Interaction of silver nanoparticles with Mediterranean agricultural soils: Lab-controlled adsorption and desorption studies. *J. Environ. Sci.*, **83**, pp. 205–216.
6. Gladkova, M. M. and Terekhova, V. A. (2013). Engineered nanomaterials in soil: Sources of entry and migration pathways. *Mosc. Univer. Soil Sci. Bull.*, **68**(3), pp. 129–134.
7. Whitley, A. R., Levard, C., Oostveen, E., Bertsch, P. M., Matocha, C. J., von der Kammer, F., and Unrine, J. M. (2013). Behavior of Ag nanoparticles in soil: Effects of particle surface coating, aging and sewage sludge amendment. *Environ. Pollut.*, **182**, pp. 141–149.
8. Rawat, S., Pullagurala, V. L., Adisa, I. O., Wang, Y., Peralta-Videa, J. R., and Gardea-Torresdey, J. L. (2018). Factors affecting fate and transport of engineered nanomaterials in terrestrial environments. *Curr. Opin. Environ. Sci. Health*, **6**, pp. 47–53.
9. Farré, M., Sanchís, J., and Barceló, D. (2011). Analysis and assessment of the occurrence, the fate and the behavior of nanomaterials in the environment. *TrAC Trends Anal. Chem.*, **30**(3), pp. 517–527.
10. Pachapur, V. L., Larios, A. D., Cledón, M., Brar, S. K., Verma, M., and Surampalli, R. Y. (2016). Behavior and characterization of titanium dioxide and silver nanoparticles in soils. *Sci Total Environ*, **563**, pp. 933–943.

11. Tourinho, P. S., Van Gestel, C. A., Lofts, S., Svendsen, C., Soares, A. M., and Loureiro, S. (2012). Metal-based nanoparticles in soil: Fate, behavior, and effects on soil invertebrates. *Environ Toxicol Chem*, **31**(8), pp. 1679–1692.

12. Giese, B., Klaessig, F., Park, B., Kaegi, R., Steinfeldt, M., Wigger, H., ... and Gottschalk, F. (2018). Risks, release and concentrations of engineered nanomaterial in the environment. *Sci. Rep.*, **8**(1), pp. 1–18.

13. Chaudhuri, R. G. and Paria, S. (2012). Core/shell nanoparticles: Classes, properties, synthesis mechanisms, characterization, and applications. *Chem. Rev.*, **112**(4), pp. 2373–2433.

14. Mahmoudi, M., Lynch, I., Ejtehadi, M. R., Monopoli, M. P., Bombelli, F. B., and Laurent, S. (2011). Protein–nanoparticle interactions: Opportunities and challenges. *Chem. Rev.*, **111**(9), pp. 5610–5637.

15. Yu, S. J., Yin, Y. G., and Liu, J. F. (2013). Silver nanoparticles in the environment. *Environ. Sci. Processes Impacts*, **15**(1), pp. 78–92.

16. EMBRAPA (2020). *Nanotecnologia*. Empresa Brasileira de Pesquisa Agropecuária. Available at:< https://www.embrapa.br/tema-nanotecnologia/nota-tecnica>.

17. Niederberger, M. and Pinna, N. (2009). *Metal Oxide Nanoparticles in Organic Solvents: Synthesis, Formation, Assembly and Application.* Springer Science & Business Media.

18. Amde, M., Liu, J. F., Tan, Z. Q., and Bekana, D. (2017). Transformation and bioavailability of metal oxide nanoparticles in aquatic and terrestrial environments. A review. *Environ. Poll.*, **230**, pp. 250–267.

19. Ju-Nam, Y. and Lead, J. (2016). Properties, sources, pathways, and fate of nanoparticles in the environment. *Eng. Nano. Environ. Biophysicochemical Process. Toxic.*, **4**, pp. 95–117.

20. Auffan, M., Rose, J., Wiesner, M. R., and Bottero, J. Y. (2009). Chemical stability of metallic nanoparticles: A parameter controlling their potential cellular toxicity in vitro. *Environ. Pollut.*, **157**(4), pp. 1127–1133.

21. Von Moos, N., Bowen, P., and Slaveykova, V. I. (2014). Bioavailability of inorganic nanoparticles to planktonic bacteria and aquatic microalgae in freshwater. *Environ. Sci.: Nano*, **1**(3), pp. 214–232.

22. Zhang, Y., Newton, B., Lewis, E., Fu, P. P., Kafoury, R., Ray, P. C., and Yu, H. (2015). Cytotoxicity of organic surface coating agents used for nanoparticles synthesis and stability. *Toxicol. In Vitro*, **29**(4), pp. 762–768.

23. Máté, Z., Horváth, E., Kozma, G., Simon, T., Kónya, Z., Paulik, E., Papp, A., and Szabó, A. (2016). Size-dependent toxicity differences of

intratracheally instilled manganese oxide nanoparticles: Conclusions of a subacute animal experiment. *Biol. Trace Elem. Res.*, **171**(1), pp. 156–166.

24. Singh, S. P., Kumari, M., Kumari, S. I., Rahman, M. F., Mahboob, M., and Grover, P. (2013). Toxicity assessment of manganese oxide micro and nanoparticles in Wistar rats after 28 days of repeated oral exposure. *J. Appl. Toxicol.*, **33**(10), pp. 1165–1179.

25. Lowry, G. V., Gregory, K. B., Apte, S. C., and Lead, J. R. (2012). Transformations of nanomaterials in the environment. *Environm. Sci. Technol.*, **46**, pp. 6893–6899.

26. Wang, D., Gao, Y., Lin, Z., Yao, Z., and Zhang, W. (2014). The joint effects on *Photobacterium phosphoreum* of metal oxide nanoparticles and their most likely coexisting chemicals in the environment. *Aquat. Toxicol.*, **154**, pp. 200–206.

27. Bayat, A. E., Junin, R., Shamshirband, S., and Chong, W. T. (2015). Transport and retention of engineered Al_2O_3, TiO_2, and SiO_2 nanoparticles through various sedimentary rocks. *Scientific Reports*, **5**, pp. 14264.

28. Adamczyk, Z. and Weroński, P. (1999). Application of the DLVO theory for particle deposition problems. *Adv. Colloid Interface Sci.*, **83**(1–3), pp. 137–226.

29. Hotze, E. M., Phenrat, T., and Lowry, G. V. (2010). Nanoparticle aggregation: Challenges to understanding transport and reactivity in the environment. *J. Environ. Qual.*, **39**(6), pp. 1909–1924.

30. Quik, J. T., Stuart, M. C., Wouterse, M., Peijnenburg, W., Hendriks, A. J., and van de Meent, D. (2012). Natural colloids are the dominant factor in the sedimentation of nanoparticles. *Environ. Toxicol. Chem.*, **31**(5), pp. 1019–1022.

31. Schultz, C., Powell, K., Crossley, A., Jurkschat, K., Kille, P., Morgan, A. J., ... and Spurgeon, D. J. (2015). Analytical approaches to support current understanding of exposure, uptake and distributions of engineered nanoparticles by aquatic and terrestrial organisms. *Ecotoxicology*, **24**(2), pp. 239–261.

32. Cornelis, G., Hund-Rinke, K., Kuhlbusch, T., Van den Brink, N., and Nickel, C. (2014). Fate and bioavailability of engineered nanoparticles in soils: A review. *Crit. Rev. Environ. Sci. Technol.*, **44**(24), pp. 2720–2764.

33. Milani, N., McLaughlin, M. J., Stacey, S. P., Kirby, J. K., Hettiarachchi, G. M., Beak, D. G., and Cornelis, G. (2012). Dissolution kinetics of macronutrient fertilizers coated with manufactured zinc oxide nanoparticles. *J. Agric. Food Chem.*, **60**(16), pp. 3991–3998.

34. Milani, N., Hettiarachchi, G. M., Kirby, J. K., Beak, D. G., Stacey, S. P., and McLaughlin, M. J. (2015). Fate of zinc oxide nanoparticles coated onto macronutrient fertilizers in an alkaline calcareous soil. *PLoS One*, **10**(5).

35. Zhao, L., Peralta-Videa, J. R., Ren, M., Varela-Ramirez, A., Li, C., Hernandez-Viezcas, J. A., ... and Gardea-Torresdey, J. L. (2012). Transport of Zn in a sandy loam soil treated with ZnO-NPs and uptake by corn plants: Electron microprobe and confocal microscopy studies. *Chem. Eng. J.*, **184,** pp. 1–8.

36. Cornelis, G., Kirby, J. K., Beak, D., Chittleborough, D., and McLaughlin, M. J. (2010). A method for determination of retention of silver and cerium oxide manufactured nanoparticles in soils. *Environ. Chem.*, **7**(3), pp. 298–308.

37. Cornelis, G., Ryan, B., McLaughlin, M. J., Kirby, J. K., Beak, D., and Chittleborough, D. (2011). Solubility and batch retention of CeO_2 nanoparticles in soils. *Environ. Sci. Technol.*, **45**(7), pp. 2777–2782.

38. Ghosh, S., Mashayekhi, H., Pan, B., Bhowmik, P., and Xing, B. (2008). Colloidal behavior of aluminum oxide nanoparticles as affected by pH and natural organic matter. *Langmuir*, **24**(21), pp. 12385–12391.

39. Romero-Freire, A., Lofts, S., Martín Peinado, F. J., and van Gestel, C. A. (2017). Effects of aging and soil properties on zinc oxide nanoparticle availability and its ecotoxicological effects to the earthworm *Eisenia andrei. Environ. Toxicol. Chem*, **36**(1), pp. 137–146.

40. Wang, P., Menzies, N. W., Lombi, E., McKenna, B. A., Johannessen, B., Glover, C. J., ... and Kopittke, P. M. (2013). Fate of ZnO nanoparticles in soils and cowpea (*Vigna unguiculata*). *Environ. Sci. Technol.*, **47**(23), pp. 13822–13830.

41. Dimkpa, C. O. (2013). Soil properties influence the response of terrestrial plants to metallic nanoparticles exposure. *Curr. Opin. Environ. Sci. Health*, **6**, pp. 1–8.

42. Du, W., Sun, Y., Ji, R., Zhu, J., Wu, J., and Guo, H. (2011). TiO_2 and ZnO nanoparticles negatively affect wheat growth and soil enzyme activities in agricultural soil. *J. Environ. Monit.*, **13**(4), pp. 822–828.

43. Bian, S. W., Mudunkotuwa, I. A., Rupasinghe, T., and Grassian, V. H. (2011). Aggregation and dissolution of 4 nm ZnO nanoparticles in aqueous environments: Influence of pH, ionic strength, size, and adsorption of humic acid. *Langmuir*, **27**(10), pp. 6059–6068.

44. Jiang, C., Aiken, G. R., and Hsu-Kim, H. (2015). Effects of natural organic matter properties on the dissolution kinetics of zinc oxide nanoparticles. *Environ. Sci. Technol.*, **49**(19), pp. 11476–11484.

45. Li, M., Lin, D., and Zhu, L. (2013). Effects of water chemistry on the dissolution of ZnO nanoparticles and their toxicity to *Escherichia coli*. *Environ. Pollut.*, **173**, pp. 97–102.

46. Miao, L., Wang, C., Hou, J., Wang, P., Ao, Y., Li, Y., ... and Xu, Y. (2015). Enhanced stability and dissolution of CuO nanoparticles by extracellular polymeric substances in aqueous environment. *J. Nanopart. Res.*, **17**(10), pp. 404.

47. Odzak, N., Kistler, D., Behra, R., and Sigg, L. (2015). Dissolution of metal and metal oxide nanoparticles under natural freshwater conditions. *Environ. Chem.*, **12**(2), pp. 138–148.

48. Wang, Z., Zhang, L., Zhao, J., and Xing, B. (2016). Environmental processes and toxicity of metallic nanoparticles in aquatic systems as affected by natural organic matter. *Environ. Sci. Nano*, **3**(2), pp. 240–255.

49. Omar, F. M., Aziz, H. A., and Stoll, S. (2014). Aggregation and disaggregation of ZnO nanoparticles: Influence of pH and adsorption of Suwannee River humic acid. *Sci. Total Environ.*, **468**, pp. 195–201.

50. Dong, C., Chen, W., and Liu, C. (2014). Preparation of novel magnetic chitosan nanoparticle and its application for removal of humic acid from aqueous solution. *Appl. Surf. Sci.*, **292**, pp. 1067–1076.

51. Shen, X. E., Shan, X. Q., Dong, D. M., Hua, X. Y., and Owens, G. (2009). Kinetics and thermodynamics of sorption of nitroaromatic compounds to as-grown and oxidized multiwalled carbon nanotubes. *J. Colloid Interface Sci.*, **330**(1), pp. 1–8.

52. Yu, S., Liu, J., Yin, Y., and Shen, M. (2018). Interactions between engineered nanoparticles and dissolved organic matter: A review on mechanisms and environmental effects. *J. Environ. Sci.*, **63**, pp. 198–217.

53. Ema, M., Kobayashi, N., Naya, M., Hanai, S., and Nakanishi, J. (2010). Reproductive and developmental toxicity studies of manufactured nanomaterials. *Reprod Toxicol.*, **30**(3), pp. 343–352.

54. Campagnolo, L., Massimiani, M., Magrini, A., Camaioni, A., and Pietroiusti, A. (2012). Physico-chemical properties mediating reproductive and developmental toxicity of engineered nanomaterials. *Curr. Med. Chem.*, **19**(26), pp. 4488–4494.

55. Sun, J., Zhang, Q., Wang, Z., and Yan, B. (2013). Effects of nanotoxicity on female reproductivity and fetal development in animal models. *Int. J. Mol. Sci.*, **14**(5), pp. 9319.

Chapter 5

Occupational Health Hazards of Nanoparticles

Sandra Magali Heberle[a] and Michele dos Santos Gomes da Rosab[b]

[a]Departamento de Fisioterapia–Universidade CESUCA, RS, Brazil
[b]Cardiovascular Centre of Universidade de Lisboa–CCUL,
Falculty of Medicine, Universidade de Lisboa, Portugal
sandra.heberle@cesuca.edu.br

5.1 Introduction

Nanoparticles are part of human life, but especially in recent years, there has been a rapid development of nanotechnologies and their application in medicine, industry, engineering, etc. (Adamec et al. 2019a). Many modern processes use nanomaterals, globalization, and global expansion of supply chains, and technological changes have an impact on both developed and developing countries. There is talk of an unrigorous form of emerging risks that include:

Environmental, Ethical, and Economical Issues of Nanotechnology
Edited by Chaudhery Mustansar Hussain and Gustavo Marques da Costa
Copyright © 2022 Jenny Stanford Publishing Pte. Ltd.
ISBN 978-981-4877-76-3 (Hardcover), 978-1-003-26185-8 (eBook)
www.jennystanford.com

technological innovations such as biotechnology/nanoparticles) (Silva and Hurt 2014). The term nanotechnology refers to the use of technologies involving the creation and manipulation of materials for the development of new materials and products of nanometric size in order to explore new characteristics more efficient, increasingly used today (Silva, Arezes, and Swuste 2015), so much so that more and more products manufactured worldwide use nanotechnologies. We are dealing with a new technology, and the health risks associated with the uses of different nanomaterials are still little known.

Nanoparticles are particularly dangerous because they can be absorbed directly into the bloodstream through the skin and through the lung membranes by inspiration. The usual personal protective equipment does not provide adequate protection. Toxic dust sits when working with substances that are themselves toxic (e.g., lead, mercury, chromio, etc.). If inhaled, they can damage the lungs or enter the bloodstream and spread throughout the body.

The term "dusts" has no exact scientific meaning but is usually defined as a solid that has been reduced to dust or fine particles. The particle size is as important as the nature of the powder to establish whether the substance is so dangerous. In general, more dangerous types of dust are those with very small particles, invisible to the human eye, as is the case with fine powders. These types of particles are small enough to be inhaled, but at the same time large enough to remain trapped in lung tissue and not be exuded.

The size of a particle is expressed in length and width values, which are generally only valid for morphology particles and regular geometry. For example, a sphere is defined by radius or diameter, while a cube is by height, length, and width. Existing methods circumvent this situation, which occurs in most nanoparticles, by estimating an average particle size based on their mass or considering nanoparticles as spheres of equivalent properties.

Each system has different requirements for particle size, but to be considered nanoparticles, they must have average particle sizes of about 100 nm (nanometers). The physical stability of the systems depends largely on the size of the particles (SÖYLEMEZ 2017).

The latest guidelines have suggested that particle weight and volume may be useful criteria for measuring air exposure of nanomaterials. Since the first publication of Ademec et al. (2019b), the government has made instructions on how to assess personal exposure and implement strategies for risk management (Adamec et al. 2019a; Hussain 2020c).

Thus, the distribution of particle sizes or dispersion of nanoparticles, depending on the measurement method, can be based on the sum of the volumes of all particles by fractional volumes of the sample.

Dispersions or suspensions of nanoparticles can be monodispersed when the distribution of particle sizes is very small, i.e., they have very low polydispersion indices. And they can be polydisperal if the index is high, which indicates that there is a high variability between the size of the particles in the dispersion.

Although animal studies have shown that these particles result in damage to the lungs of rats, there has been growing concern that long-standing exposure to nanoparticles without protective measures may be related to serious lung damage in humans.

There are two types of sources of emission or production of nanoparticles: natural origin (volcanic eruptions, forest fires, marine pollution) and of unintentional anthropogenic origin (industrial pollution, diesel emissions, various combustions, indoor pollution inside buildings) and other intentional (nanoparticles manufactured on an industrial scale or on the laboratory scale related to the development industry (R&D)) (Jasti et al. 2008; Hussain 2018; Hussain 2020a; Hussain 2020b).

The chemical nature of nanoparticles can be more or less complex, of mineral origin (graphite, hematite, silica); metallic or organic: carbon-fullerene compounds, single-walled carbon nanotubes (SWCNTs), and multi-walled carbon nanotubes (MWCNTs); polymers, nylon, dextrane, polystyrene, etc. Still its degree of complexity may be associated with the generation process (heating of polytetrafluoroetieleno (PTFE) or Teflon, welding fumes, origin in the combustion of a hydrocarbon or a polymer). And they can be composed of a nanomaterial that serves as a nucleus, to which pollutants (transition metals, hydrocarbons, or biological substances) are absorbed (Jasti et al. 2008).

Annoying dusts, these can be generated by handling materials such as:

- Flour
- Tobacco
- Sugar
- Paper
- Dry foodstuff
- Cement
- Sawdust
- Coffee beans and tea
- Carbon black (toner for photocopier/printer)

These types of dust are generally only irritating, but concentrated form can be dangerous to health, such as hardwood powder that is carcinogenic (Tonet et al. 2019).

The most important diseases associated with inhalation of hazardous dust are: Benign penumoconiosis caused when seemingly harmless dust is inhaled and deposited in the lungs in such a way that it becomes visible through X-rays. This is a disturbance most often associated with dust from metals such as iron and tin.

Pneumoconiosis is a collective name for a group of chronic lung disease caused by inhalation of poerias containing specific minerals. The term includes a number of diseases whose names come from the dusts that cause them.

The best known are:

- Asbestosis (asbestos dust)
- Silicosis (silica dust)
- Talcose (talcum dust)

Pneumonitis is the inflammation of human tissues or bronchioles essentially caused by inhalation of metal-containing dust. Symptoms are similar to pneumonia, but the severity level varies depending on the inhaled metal. The most common causes are cadmium and beryllium dust.

Mesothelioma of the pleura is a tumor of the lungs, mainly caused by exposure to asbestos. Lung cancer can be followed by any exposure to asbestos.

The risks of exposure to nanomaterials are linked to the three potential exposure pathways: inhalation, ingestion, and dermal contact. The respiratory system is the main pathway of penetration of nanoparticles into the human body, being this pathway the most important in individuals who present alterations or decreased lung capacity. Nanoparticles, once inhaled, can be deposited in different regions of the respiratory system. This deposition is not uniform throughout the respiratory system, varies depending on the diameter, degree of aggregation and agglomeration, and air-borne behavior of these particles. Also, they can also be found in the gastrointestinal system, after being ingested or swallowed or inhaled (Silva, Arezes, and Swuste 2015). Once in the body, these particles can move to different regions or organs far from the entrance.

There is a consensus that the greatest risk of nanoparticles to a human organism is due to their inhalation. Clinical studies with humans, rats, and lung cell culture were conducted, and in most of these, some toxic effect was observed, such as lung inflammation, asthma, chronic pulmonary obstruction, or even death, and it is estimated that 50 to 80% of human exposure to inhalable nanoparticles is due to various internal sources. The main concern with inhalation of nanoparticles is that the smaller the particle, the more easily it overcomes the natural barriers of the respiratory system, being deposited and accumulated in the alveoli, responsible for the gas exchange of O_2 and CO_2 in the bloodstream @article (Paschoalino, Marcone, and Jardim 2010).

The health risks arising from nanoparticles in the workplace may be significantly higher than those related to the same pollutants in domestic occurrences, since exposure concentrations in such cases may be substantially higher. Lung is vulnerable to exposure of a wide range of harmful substances in the workplace, including allergens, asthma, organic powders, dust and mineral fibers, solvents, gases, and carcinogenic chemicals (Holgate et al. 2016).

In the case of people who are working for several consecutive hours, within environments, where the internal quality of the air is different from the external, some effects on health may appear after repeated exposures to a particular pollutant or substance. These effects include irritation of the eyes, airways, and throat, headaches, dizziness, and generalized tiredness, which may be short lived and treatable. Soon after prolonged exposure to some internal pollutants, symptoms of some diseases, such as asthma, may appear aggravated (Arbex et al. 2007).

The likelihood of immediate reactions to indoor air pollutants depends on several factors, including age and pre-existing health conditions. In some cases, if a person reacts to a pollutant, the effect will depend on individual sensitivity, which varies tremendously from individual to individual. Some people may become sensitized to biological or chemical pollutants after repeated or high-level exposures (Guieysse et al. 2008).

Certain immediate effects are similar to those of colds or other viral diseases. Therefore, it is often difficult to determine precisely whether the symptoms are the result of exposure to indoor air pollution, or whether the cause is another. Therefore, it is important to pay attention to the time and place where symptoms occur. Some effects may be worsened due to the possible improper supply of outside air that comes into the vehicle, or to prevailing heating, cooling, or humidity conditions indoors (Heberle et al. 2018).

Thus, the effects of nanomaterials that cause greater concern occur in the lungs, among others, inflammation and tissue lesions, the appearance of fibrosis and tumors. The cardiovascular system may also be affected. For example, some types of carbon nanotubes can have the same effects as asbestos (Muhlfeld et al. 2008).

Substances can deposit in different regions of the respiratory tract, and this is what can trigger different symptoms. Once deposited in the pulmonary epithelium, in contrast to larger particles, the nanoparticles can translocate to extrapulmonary sites thus reaching other organs by different routes and mechanisms. A possible mechanism would be transcytosis through the epithelium of the respiratory tract accessing the bloodstream directly or transported by lymphocytes, resulting in the distribution of nanoparticles throughout the body.

5.2   Anatomy of the Respiratory System

The respiratory system consists of the lungs and various organs that allow the air to travel into and out of the pulmonary structures. These organs are the nasal fossas, the mouth, the pharynx, the larynx, the trachea, the bronchi, the bronchioles, and the alveoli (Gray and NETTER 1949).

Chest: The lungs are located inside the chest. The ribs, which form the rib cage and make a framework protecting the lungs and heart, lean forward by the action of the intercostal muscle, causing an increase in the volume of the thoracic cavity. The volume of the chest also increases by contraction down the diaphragm muscles. When the thorax expands, during inspiration, the lungs begin to fill with air (Adamec et al. 2019b).

Human lungs are spongy organs approximately 25 cm long, being surrounded by a membrane called pleura, which, after lining them, will line the inner wall of the thoracic cavity. Thus, a visceral pleura (involving the lungs) and a parietal pleura (which adheres to the rib cage) are distinguished. In the lungs, the bronchi branch out and give rise to bronchioles, which are thinner units (Gray and NETTER 1949).

The highly branched set of bronchioles is the bronchial tree or respiratory tree. Each bronchiolo ends in small pockets formed by flattened epithelial cells, covered by blood capillaries, called pulmonary alveoli.

The lung rests on the diaphragm, a muscle that separates the thorax from the abdomen, promoting, along with the intercostal muscles, respiratory movements. Located just above the stomach, the phrenic nerve controls diaphragm movements (Adamec et al. 2019b).

The most characteristic property of the lung is its elasticity: The air that penetrates it forces it to dilate, returning to its primitive volume when that air comes out.

5.2.1 Respiratory Control Breathing

Respiratory control breathing is automatically controlled by a nerve center located in the bulb. From this center, the nerves responsible for the contraction of the respiratory muscles (diaphragm and intercostal muscles) depart. Nerve signals are transmitted from this point, through the spinal column, to the muscles of breathing. The diaphragm receives respiratory signals through the phrenic nerve, which leaves the spinal cord in the upper half of the neck, heading down, through the chest, to the diaphragm (Gray and NETTER 1949).

Signals for expiratory muscles, especially abdominal muscles, are transmitted to the lower portion of the spinal cord and to the spinal nerves that inflate the muscles. Under normal conditions, the respiratory center (CR) produces, every 5 sec, a nerve impulse that stimulates the contraction of the thoracic musculature and diaphragm, making the inspiration.

The CR increases and decreases both the frequency and amplitude of respiratory incursions, as it has chemoreceptors that are very sensitive to plasma pH, which allows tissues to receive the amount of oxygen needed and, in addition, adequately removes carbon dioxide. When blood becomes more acidic due to increased carbon dioxide and pollutants, CR induces respiratory rate acceleration. Thus, both the frequency and amplitude of breathing become increased due to the excitation of the CR. In a contrary situation, with CR depression, there is a decrease in respiratory frequency and amplitude.

5.2.2 Lung Capacity and Lung Volumes

The human respiratory system holds a total volume of approximately 5800 ml of air, which is called total lung capacity (TLC). Of this volume, only half a liter is renewed in each breath at rest. Such renewed volume is the tidal volume. If, at the end of a forced inspiration, a forced exhalation is performed, an amount of approximately 4 L of air will be removed from the lungs, which corresponds to the vital capacity (CV), and it is within its limits that breathing can happen. Even at the end of a forced expiration, about 1 L of air remains in the airways, called residual volume (VR) (Costa and Jamami 2001).

In the evaluation of ventilation and lung capacity, the following lung volumes are considered: tidal volume (VC), inspiratory reserve volume (IVR), expiratory reserve volume (VRE), and residual volume (RV) (Costa and Jamami 2001).

These volumes, shown in Fig. 2.4, can be described as follows: Tidal volume is the volume of air inspired or exhaled spontaneously in each respiratory cycle (500 ml); inspiratory reserve volume is the maximum extra volume that can be voluntarily inspired at the end of a forced maximum inspiration; expiratory reserve volume is the maximum volume that can be voluntarily expired at the end of a spontaneous expiration of the tidal volume; residual volume is the volume of gas that remains inside the lungs after maximum expiration. When lung function is evaluated, the main lung capacities examined are as follows:

- Inspiratory Capacity (IC): It is the maximum volume that can be voluntarily inspired from the end of a spontaneous inspiration, the sum of VC and VRI.
- Functional Residual Capacity (CRF): It is the volume of air that remains in the lungs at the end of a normal expiration, corresponding to the sum of VRE and VR.
- Vital Capacity (CV): It is the amount of air mobilized in the maximum inspiration/expiration, being the sum of VC, VRI, and VRE.
- Total Lung Capacity (TLC): It is the amount of air in the lungs at the end of a maximum inspiration, being, therefore, the sum of tidal, inspiratory reserve, expiratory reserve, and residual volumes (Holgate et al. 2016).

5.2.3   Spirometry

The term spirometry comes from Latin (spiro = breathe and metrum = measured), and the spirometry test, therefore, consists of measuring the intake and outlet of air in the lungs (Costa and Jamami 2001).

This is an examination used as a test of pulmonary function, whose main objectives are the following: detect and prevent obstructive and/or restrictive pulmonary changes; differentiate a functional obstructive disease from an organic obstructive disease; evaluate the clinical course of a respiratory disease; assess surgical risk; direct treatment approaches in patients with heart disease; subsidize the assessment of workers' health, especially in the control of industrial risks.

The values for pulmonary function differ substantially between the various regions of the world, which has been attributed to anthropometric, environmental, social, and genetic factors, as well as technical factors. Attempts to compile equations from different authors were made for Europe in 1983 and then in 1993. Such recommendations of the working group were accepted and made official by the European Respiratory Society (ERS), with the recommendation for their widespread use in Europe (Holgate et al. 2016).

Spirometry can be performed with many different types of equipment. It requires cooperation between the subject and the examiner, and the results obtained will depend on technical and personal factors. All spirometers must follow the quality standards approved by the American Thoracic Society (ATS) or the British Thoracic Society (BTS) and, in Brazil, by the National Health Surveillance Agency.

In short, a spirometry examination can be obtained with a very simple maneuver: It is deeply inspired, until it fully fills the lungs (total lung capacity), and at a prolonged expiration, the entire volume of air contained in the lungs is blown for as long as possible (at least 6 sec). Thus, the necessary parameters are obtained, especially the forced vital capacity (FVC), which is the maximum volume of air eliminated, and the forced expiratory volume in the first second (FEV1) (Czubacka and Czerczak 2019).

The values of the spirometry test results should be expressed in volume–time and flow–volume graphs (Fig. 5.1). It is essential that a graphic record follows the numerical values obtained in the test (Holgate et al. 2016), because they are important for the control of their quality in the evaluation of the magnitude of the patient's effort at the beginning of the examination and also to show the end of the measurement and the duration of the effort (Costa and Jamami 2001).

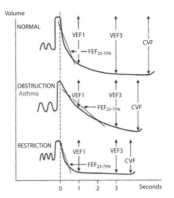

Pulmonary Ventilation

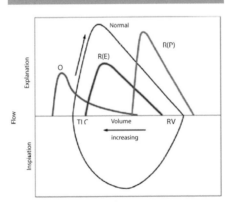

Spirometry

Figure 5.1 Comparison of flow-volume and volume-time curves presenting normal, obstructive and restrictive patterns. F: Flow, V: Volume, t: Time, FVC: Forced Vital Capacity, FEV: Forced Expiratory Volume, FeF: Forced Expiratory Flow, R: Restrictive; O: Obstructive.

The complexity of the examination is due to three important points: determination of the predicted value, multiple parameters evaluated during its performance, and the determination of the type of disorder. There are several parameters evaluated; however, after some studies, it was determined that it is possible to obtain a satisfactory result, using only these three parameters: forced vital capacity (FVC), forced expiratory volume in the first second (FEV1), and Tiffeneau index, which occurs by fev1/FVC ratio (Holgate et al. 2016).

For the number of attempts to perform the test, it should be taken into account that three acceptable and two reproducible curves are necessary, that there is a learning effect with the repetition of maneuvers, that there is the possibility of an individual getting tired or bored, and that, rarely, more than eight curves would be necessary (Holgate et al. 2016).

However, if the reproducibility criteria are not met, new examination maneuvers should be obtained so that it can be repeated as many times as necessary. It is noteworthy, however, that if reproducible values are not reached after eight attempts, the procedure should be suspended and repeated on another day.

5.2.4 Basic Standards to be Recognized in Spirometry

There are three basic standards to be recognized in spirometry: Normal— FEV1 and FVC above 80% of predicted; FEV1/FVC ratio above 0.7. Obstructive--any process that interferes with the airflow of the lungs can be considered as obstruction in the airways: FEV1 below 80% of the predicted; FVC may be normal or reduced, usually to a degree lower than FEV1; FEV1/FVC ratio less than 0.7. Restrictive–there is a reduction in total lung capacity: FEV1 and FVC below 80% of predicted; normal FEV1/FVC ratio or above 0.7.

Exposure assessment: Recent studies have shown that chronic exposure to indoor air pollutant levels, including nanoparticles, can have even higher impacts on lung capacity, respiratory diseases, and mortality than acute exposure (Cançado et al. 2006).

It is known that the occupational environment has an influence on the pulmonary function of individuals, and there are reports of latency time. On average, it is 10 to 15 years of exposure to toxic residues to present respiratory problems. This delay time is a consequence of the fact that the changes start through the distal airways, due to the size of the particles of pollutants, so that the tendency is that changes in pulmonary function present as the disease evolves, and the most common form is obstructive pulmonary disease (Cançado et al. 2006).

A major Lancet publication in 2014 highlighted that gases generated from traffic and energy use are the main sources of urban air pollution. The idea that outdoor air pollution can reach lung capacity and cause exacerbations of pre-existing asthma is supported by a base of evidence that is being accumulated for several decades, with several other studies suggesting a contribution to health research. The article discusses the effects of air pollution related to mixed traffic, polluting gases (ozone, NO_2, and sulfur dioxide), and particulate matter. It focuses on clinical, epidemiological, and experimental studies published in the last 5 years and, from a mechanistic perspective, declares that air pollutants probably cause oxidative injury of the airways, leading to inflammation, remodeling, and increased risk of sensitization.

Although several pollutants have been associated with asthma, the strength of evidence varies. In a study prepared in Canada, air quality was classified into six categories: good, when the concentrations of all pollutants are below 50%; concentration of at least one of the pollutants reaches its quality standard; inadequate, when the concentration of at least one of the pollutants is between its quality standard and the levels of care; concentration of at least one of the pollutants is among their levels of attention and alertness; concentration of at least one of the pollutants is among their alert and emergency levels (Crouse et al. 2015).

In accordance with several studies that seek the aggravating effects of nanoparticles on human health, it is suggested that, although concentration during exposure is considered low to cause some type of important clinical symptom in the long term, it can lead to biochemical and pulmonary changes that cause chronic lung diseases (Organization et al. 2017).

The evaluation of occupational exposure to nanoparticles is similar to the methodology used for professional evaluation of chemical substances in the workplace: analysis of the job, types of sources, and exposure time. To establish environmental exposure limit values, it is necessary to define the relationships between exposure and health effects. Therefore, environmental monitoring and biological monitoring should be carried out when there is a biological indicator of exposure to the parent substance. Therefore, there is the need to use biomarkers to improve early biochemical and molecular changes, establish mechanisms of action to explain the final results of research on respiratory, cardiovascular function, immune response, and possible development of cancer associated with exposure.

The effects on human health, resulting from exposure to environmental pollution, are many of different intensities and manifest themselves with different latency times (Heberle et al. 2018).

The main ones are as follows: behavioral and cognitive effects, pulmonary and systemic diseases, changes in airway caliber, cardiovascular changes, reproductive changes, and morbidity and mortality due to cardiorespiratory diseases and increased incidence of neoplasms. From these findings, it is possible to select the useful events to determine the impact that some change in the environment will have on the population that is exposed (Saldiva et al. 2010).

The concept most used to demonstrate an adverse effect on people's health has been that recommended by the American Thoracic Society (1995), describing the health problem "as a significant medical event, characterized by one or more of the following factors: (1) interference with the normal activity of affected individuals; (2) episodic respiratory disease; (3) disabling disease; (4) permanent respiratory disease; (5) Progressive respiratory dysfunction" (Braga et al. 2001).

Even if this definition of adverse or harmful effect on human health is widely used to define forms of risk or environmental assessment, an exact definition of the acceptable values existing between statistically significant findings and changes that cause damage to health still need further clarification (Braga et al. 2001; Saldiva et al. 2010).

Many of these substances are highly soluble in tissues and, because of this, react not only with the alveolar epithelium, but also with the interstitium and endothelium of pulmonary capillaries, being an important mechanism responsible for lung lesions. Thus, much of what is breathed is retained in the lungs, being deposited in the bronchial tree and alveoli, which are the terminal respiratory units. As a result, there are lesions in the lung, through oxidizing properties or, indirectly, by increasing pulmonary fragility and being susceptible to respiratory infections (Kuriyama, Moreira, and Silva 1997).

Human beings are highly exposed to several types of substances that can impair pulmonary function and, consequently, cause serious damage to the health of the population, because air quality interferes directly in the respiratory system, due to its large contact area, and can also reach systemic circulation through the lungs, causing harmful effects on organs and systems (Brook et al. 2010).

Among the various factors that can generate the onset of pulmonary or cardiovascular diseases, nanoparticles, air pollution, and smoking are two important causes that have been studied. Not only health researchers but also scholars from other areas are concerned with exploring this relationship with the development of respiratory diseases, in order to seek prevention measures (Arbex et al. 2007).

The World Health Organization (WHO) thinks that the most effective way to eliminate the risks of nanoparticles in the workplace is personal training in the area of health risks, and to practice procedures for personal protection. In addition to training, the use of personal protective equipment, such as protective masks, is also important (Organization et al. 2017).

The European Commission also recommends taking operational action (European Commission, 2014):

- Reduce the number of employees who come into contact with nanoparticles.

- Reduce the time that employees who come into contact with nanoparticles remain exposed.

- Change the current working procedures. Introduce stricter health procedures (Castillo 2019).

5.3 Nanomedicine and Pulmonary Disease Therapy

The use of nanotechnology has increased significantly in several areas of science. They are the development of drug release systems. Today, the most modern pharmaceutical nanocarriers, such as liposomes, micelles, nanoemulsions, and polymeric nanoparticillans, demonstrate extremely useful properties from the pharmacotherapeutic point of view. Nanocarriers for pulmonary application have been a topic discussed by the scientific community in recent decades, both for the local treatment of respiratory diseases and for the systemic administration of lipophilic drugs, which have a low bioavailability when administered orally (Torchilin 2008).

5.4 Nanoparticles Benefits in Nasal and Inhalation Therapy

Although investigations on the development of pulmonary-administered lipid nanoparticles are still in the early stages, it is an easy-to-access, non-invasive pathway; the epithelial barrier is thin, very irrigated, and with little metabolic activity, composing with other pathways (Todoroff and Vanbever 2011).

The nanoparticles used in this pathway must have a small diameter that prevents their elimination with expiration and allows their deposition throughout the respiratory tract area. The mucoadesivity of particles must be appropriate and is influenced by the charge of particles, their composition, size, and porosity. The fact that they are small facilitates their porosity and rapid internalization by endocytosis. Cationic nanoparticles more easily bind to the mucosa of the respiratory epithelium (Dombu and Betbeder 2013).

Eyelashes and mucus in the respiratory tract can be factors that affect the efficacy by this route, as well as phagocytosis by alveolar macrophages. Even so, this pathway is promising for the administration of nanoparticulates loaded with proteins (Dombu and Betbeder 2013).

Several inhalant nanopharmaceuticals are being developed: budesonide, salbutamol, itraconazole, and paclitaxel, both in the testing phase, targeting the treatment of pulmonary infectious diseases, and in sustained improvement of pulmonary function, and significant reduction in bacterial density in patients with cystic fibrosis (Zhang et al. 2011).

References

Adamec, Vladimír, Barbora Shüllerová, Klaudia Köbölová, Stela Pavlíková, Danuše Procházková, and Tomáš Zeman. 2019a. "Current Approaches of Occupational and Safety Health Management in Work Environments Containing Nanoparticles."

———. 2019b. "Current Approaches of Occupational and Safety Health Management in Work Environments Containing Nanoparticles."

Arbex, Marcos Abdo, Lourdes Conceicao Martins, Luiz Alberto Amador Pereira, Fernanda Negrini, Arnaldo Alves Cardoso, Wanessa Roberto Melchert, RF Arbex, Paulo Hilário Nascimento Saldiva, Antonella Zanobetti, and Alfesio Luis Ferreira Braga. 2007. "Indoor No2 Air Pollution and Lung Function of Professional Cooks."*Brazilian Journal of Medical and Biological Research* **40** (4): 527–34.

Braga, Alfesio, Luiz Alberto Amador Pereira, György Miklós Böhm, and Paulo Saldiva. 2001. "Poluição Atmosférica e Saúde Humana." *Revista USP*, no. **51**: 58–71.

Brook, Robert D, Sanjay Rajagopalan, C Arden Pope III, Jeffrey R Brook, Aruni Bhatnagar, Ana V Diez-Roux, Fernando Holguin, et al. 2010. "Particulate Matter Air Pollution and Cardiovascular Disease: An Update to the Scientific Statement from the American Heart Association." *Circulation* **121** (21): 2331–78.

Cançado, José Eduardo Delfini, Alfesio Braga, Luiz Alberto Amador Pereira, Marcos Abdo Arbex, Paulo Hilário Nascimento Saldiva, and Ubiratan de Paula Santos. 2006. "Repercussões Clínicas Da Exposição à Poluição Atmosférica." *Jornal Brasileiro de Pneumologia* **32**: S5–11.

Castillo, Aida Ponce Del. 2019. "Training for Workers and Safety Representatives on Manufactured Nanomaterials." *NEW SOLUTIONS: A Journal of Environmental and Occupational Health Policy* 29 (1): 36–52.

Costa, Dirceu, and Mauricio Jamami. 2001. "Bases Fundamentais Da Espirometria." *Rev Bras Fisioter* **5** (2): 95–102.

Crouse, Dan L, Paul A Peters, Perry Hystad, Jeffrey R Brook, Aaron van Donkelaar, Randall V Martin, Paul J Villeneuve, et al. 2015. "Ambient Pm2. 5, O3, and No2 Exposures and Associations with Mortality over 16 Years of Follow-up in the Canadian Census Health and Environment

Cohort (CanCHEC)." *Environmental Health Perspectives* **123** (11): 1180–86.

Czubacka, Ewelina, and Sławomir Czerczak. 2019. "Are Platinum Nanoparticles Safe to Human Health?"

Dombu, Christophe Y, and Didier Betbeder. 2013. "Airway Delivery of Peptides and Proteins Using Nanoparticles." *Biomaterials* **34** (2): 516–25.

Gray, Henry, and Frank H NETTER. 1949. "Anatomía Humana." *Ed. Salvat, Barcelona, España.*

Guieysse, Benoit, Cecile Hort, Vincent Platel, Raul Munoz, Michel Ondarts, and Sergio Revah. 2008. "Biological Treatment of Indoor Air for VOC Removal: Potential and Challenges." *Biotechnology Advances* **26** (5): 398–410.

Heberle, Sandra Magali, Gustavo Marques da Costa, Nelson Barros, and Michele SG Rosa. 2018. "The Effects of Atmospheric Pollution in Respiratory Health." In *Handbook of Environmental Materials Management*, 1–16. Springer International Publishing Cham.

Holgate, Stephen, Jonathan Grigg, Raymond Agius, John R Ashton, Paul Cullinan, Karen Exley, David Fishwick, et al. 2016. "Every Breath We Take: The Lifelong Impact of Air Pollution, Report of a Working Party." In. Royal College of Physicians.

Hussain, Chaudhery Mustansar. 2018. *Handbook of Nanomaterials for Industrial Applications.* Elsevier.

———. 2020a. *Handbook of Functionalized Nanomaterials for Industrial Applications.* Elsevier.

———. 2020b. *Handbook of Manufacturing Applications of Nanomaterials.* Elsevier.

———. 2020c. *The ELSI Handbook of Nanotechnology: Risk, Safety, ELSI and Commercialization.* John Wiley & Sons.

Jasti, Ramesh, Joydeep Bhattacharjee, Jeffrey B Neaton, and Carolyn R Bertozzi. 2008. "Synthesis, Characterization, and Theory of [9]-,[12]-, and [18] Cycloparaphenylene: Carbon Nanohoop Structures." *Journal of the American Chemical Society* **130** (52): 17646–47.

Kuriyama, Gisele Sayuri, Josino Costa Moreira, and Célia Regina Sousa da Silva. 1997. "Exposição Ocupacional Ao Dióxido de Nitrogênio (No2) Em Policiais de Trânsito Na Cidade Do Rio de Janeiro."*Cadernos de Saude Publica* **13** (4): 677–83.

Muhlfeld, Christian, Barbara Rothen-Rutishauser, Fabian Blank, Dimitri Vanhecke, Matthias Ochs, and Peter Gehr. 2008. "Interactions of

Nanoparticles with Pulmonary Structures and Cellular Responses." *American Journal of Physiology-Lung Cellular and Molecular Physiology* **294** (5): L817–29.

Organization, World Health et al. 2017. *WHO Guidelines on Protecting Workers from Potential Risks of Manufactured Nanomaterials*. World Health Organization.

Paschoalino, Matheus P, Glauciene PS Marcone, and Wilson F Jardim. 2010. "Os Nanomateriais e a Questão Ambiental." *Química Nova* **33** (2): 421–30.

Saldiva, PHN, M de F ANDRADE, SGEK MIRAGLIA, and PA de ANDRÉ. 2010. "O Etanol e a Saúde." *Etanol e Bioeletricidade: A Cana-de-açúcar No Future Da Matriz Energética. São Paulo: Luc*, 226–59.

Silva, F, P Arezes, and P Swuste. 2015. "Systematic Design Analysis and Risk Management on Engineered Nanoparticles Occupational Exposure."

Silva, Guilherme Frederico B Lenz e, and Robert Hurt. 2014. "RISK ASSESSMENT OF NANOCARBONS: USE OF ANALYTICAL HIERARCHY AND CONTROL BANDING APROACHES FOR SAFETY MANAGEMENT."

SÖYLEMEZ, A ERNUR. 2017. "YENİ NESİl FONKSİYONEL KOPOLİMER-RAFT AJAN/ORGANO-sİlİKAT NANOYAPILARININ SENTEZİ VE KARAKTERİZASYONU."

Todoroff, Julie, and Rita Vanbever. 2011. "Fate of Nanomedicines in the Lungs." *Current Opinion in Colloid & Interface Science* **16** (3): 246–54.

Tonet, Camila, Rogério de Menezes Chultz, Melissa Freire Zimmer, and Nilton Oliveira Silva. 2019. "RELAÇÃo ENTRE PNEUMOCONIOSES e o cÂNCER DE PULMÃo." *REVISTA UNINGÁ* **56** (4): 177–86.

Torchilin, Vladimir. 2008. "Multifunctional Pharmaceutical Nanocarriers: Development of the Concept." In *Multifunctional Pharmaceutical Nanocarriers*, 1–32. Springer.

Zhang, Jian, Libo Wu, Hak-Kim Chan, and Wiwik Watanabe. 2011. "Formation, Characterization, and Fate of Inhaled Drug Nanoparticles." *Advanced Drug Delivery Reviews* **63** (6): 441–55.

Chapter 6

Ethical Issues in Nanotechnology-I

Maurício Machado da Rosa

Cardiovascular Centre of Universidade de Lisboa–CCUL,
Falculty of Medicine, Universidade de Lisboa, Portugal
prof.mauriciomr@gmail.com

This chapter will start with the measures and indicators of the societal impacts of the uses of nanotechnology, through a review of the actual technologies and an evaluation of the social modifications induced by these technologies. After that we must reflect if these modifications are good or not so good and the benefits they bring together.

The equilibrium between the potential benefits and risks of nanotechnology is discussed based on judgments expressed by leading industry, academe, and government, and the results impact the economic and commercialization, social and technological convergence, quality of life, ethics and law, and the future education.

Environmental, Ethical, and Economical Issues of Nanotechnology
Edited by Chaudhery Mustansar Hussain and Gustavo Marques da Costa
Copyright © 2022 Jenny Stanford Publishing Pte. Ltd.
ISBN 978-981-4877-76-3 (Hardcover), 978-1-003-26185-8 (eBook)
www.jennystanford.com

The social acceptance of technology is prevalent throughout both scholarly and social science knowledge and could be, in certain point, an obscure idea, because it depends on an empirical content, so that societal implications of nanoscience and nanotechnology suggest a normative judgment in a way that involves inherently contentious characterizations of the values of the society.

In addition to the acceptance, which is not the only factor to be considered, the identification of ethical issues in nanotechnology must be done, and it could be called like nanoethics, and it is a hard task to execute because it influences in the way of life of the human being that will use that technology.

Some questions regarding geographical, economic, psycho-social, affective, cognitive, technical administrative, and political parameters should be considered to decide how the nanotechnology and nanoscience will impact, and an ethical discussion should consider these parameters to validate the inclusion of new nanomaterials, or other product outcome from nanoscience.

There is no agreement on how the ethical acceptability should proceed, but it must consider a large list of issues and situations that must be analyzed before and should result from a large discussion about all implications. On the other hand, we have religious and cultural issues that must be counted, and in most cases, there are differences in ethical issues. They are not global and depends directly on the region in which they are applied. For example, in some cultures/religions, a person is not permitted to take medicine or undergo a transplantation.

Some studies point to that for the discussion about the religious and cultural influences in technology, nanotechnology, and science and nanoscience acceptability, we must consider all rational points of view and measure all the influences of each one. It is not easy to measure how a new discovery will impact the society. Some impacts that are expected could not happen because the society modifies itself and the way of using some technologies to adapt itself and give the best use of a product.

6.1 Measures and Indicators of Societal Impacts

A recent study at the University of Sheffield (UK) investigates the social and economic challenges of nanotechnology. In the project, social and natural scientists worked together to provide a careful assessment of the emerging new science of nanotechnology. In addition, the role social science can play in nanotechnology's development is discussed. Nanotechnology is seen as an opportunity to investigate broader themes, such as an evaluation of the drivers behind the technology development process and how society deals with risks under uncertainty (Wood, Geldart, and Jones 2003).

Environmental indicator could give us clues about the societal impacts of something new like a nanotechnology product that could modify the pollution and health condition. An example of an environmental indicator is the air quality measured by nitrogen oxide (NO_2), monoxide of carbon (CO) and dioxide of carbon (CO_2), and particulate material (PM_{10}, $PM_{2.5}$ and PM_1)(Panis et al. 2011) (Heberle et al. 2018) (Wallington, Sullivan, and Hurley 2008) (Hussain 2020c).

An economic indicator is additional public expenditure on health and education relative to a previous year; a social indicator could be the number of new health centers introduced to a region, and an institutional one could be the number of new worker groups established after technology introduction compared to the number before (Riley 2001).

The effects of nanotechnology are expected to stimulate improvements in work efficiency in almost all sectors involving material work. Major national efforts should be focused on how nanotechnology can improve efficiency in manufacturing and energy resources utilization, reduce environmental impacts of industry and transportation, enhance healthcare, produce better pharmaceuticals, improve agriculture and food production, and expand the capabilities of information technologies. A programmatic approach should be developed to prepare productive units and to increase synergy in nanotechnology development by creating partnerships earlier in the research and development processes between venture capitalists, industry, academia, national laboratories, and funding agencies. Social scientists, economists, and public exponents must be involved from the beginning of major projects, in order to ensure equitable distribution of contributions and benefits, and to improve potential applications so they can better serve people and thus better achieve success in the market (Roco and Bainbridge 2005) (Hussain 2018) (Hussain 2020a) (Hussain 2020b).

Much of the impact of nanotechnology will occur through its convergence with other fields, especially biotechnology, information technology, and new technologies based on cognitive science. The unity of nature at the nanoscale provides the fundamental basis for the unification of science because many structures essential to life, computation, and communication and to human thought itself are based on the phenomena that take place at this scale. Maximizing human benefit will require the development of transforming tools that can be shared across fields, such as new scientific instrumentation, overarching theoretical concepts, methods of interdisciplinary communication, and fresh techniques of production such as those bridging the gap between the organic and inorganic. Technological convergence is the wave of the future, but it cannot properly transform science and technology without the investment of considerable effort to achieve maximum human benefit. These technologies are likely to bring about and require fundamental change generating new science, new technologies, new industries, new manufacturing processes and capabilities, new services, and new skills and knowledge. They also have the potential to disrupt existing industries and the current balance between societal institutions, so comprehensive convergence will require anticipatory measures (Roco and Bainbridge 2005).

Nanoscience has been growing during the last few years. One of the good indicators is the number of publications in scientific magazines and journals. Another good indicator is the impact factor of these publications, which is also growing with the first.

The impacts of nanoscience and nanotechnology are related to the society, its habits, culture, religion, influences, the perspectives of the future, environment, and other variables that might variate from one region to other. To measure impact, we might take in count the same characteristics of these variables, analyze what that technology or knowledge could change these characteristics, and after start of uses visualize the fluctuation on modifications in these variables. It sounds simple, but part of this process sometimes ends in negligence and the real impact is estimated and not measured.

If nanotechnology, as so many other technologies before, becomes a proxy on which the modernism/traditionalism conflict is debated in developing countries, that will radically affect the perception of ethical issues of nanotechnology there (Schummer 2006).

6.2 Societal Implications of Nanoscience and Nanotechnology

A recent study at the University of Sheffield (UK) investigated the social and economic challenges of nanotechnology. In the project, social and natural scientists worked together to provide a careful assessment of the emerging new science of nanotechnology. In addition, the role social science can play in nanotechnology's development is discussed. Nanotechnology is seen as an opportunity to investigate broader themes, such as evaluation of the drivers behind the technology development process and how society deals with risks under uncertainty (Wood, Geldart, and Jones 2003).

Nanotechnology is being heralded as a new technological revolution; some influences will be positive, while others see more uncommon implications, one so profound that it will touch all aspects of human society. The definitions of nanotechnology are not always clear or indeed agreed upon. Some concepts are a bit fuzzy; these derive from the importance at small length scales of physical phenomena that are less obvious for larger objects.

Debates on the social implications of nanotechnology, not on the relatively mundane applications, but on the longer-term possibilities of nanotechnology, are more drastic and stronger for society. Current applications of nanotechnology derivatives from material science and colloid technology produce incremental advances in well-established branches of applied science (Wood, Geldart, and Jones 2003).

On one side, the long-term applications are currently only anticipated in the laboratory; on the other side, if medium term comes from the continuous technological progress, this anticipates a degree of control over matter on the nanoscale that permits fabrication from a molecular level, which results in a great number of possibilities for nano-computation devices, food, pharmacy, cosmetics, health, and medicine (Wood, Geldart, and Jones 2003).

Debates between the short-term outcomes that imply on the rapid growth of nanotechnology and the implications on long-term environmental impact, which tries to slow down the growth, are important to delimitate the future. The rapid growth promises a lot of benefits and economical rentability, and the divergences between rapid or slow growth must imply civil society, media, non-governmental organizations, and the scientist, because the impacts could affect all and must be discussed and underhanded by all.

Different aspects relevant to nanoscience and nanotechnology need to be woven into the nanodiscourse, and thus, we are looking into these specific aspects with focus on the developing world. We try to provide a comprehensive framework on issues relevant to nanoscience and technology and map the social, economic, political, legal, and ethical aspects, which are applicable to the developed as well as the developing world (Bürgi and Pradeep 2006).

Their potential impacts will be stronger, because nanoscience has the incomparable force to pervade all societies and economies, from the pre-industrial to knowledge societies, from ancestral to highly industrialized economies and is not necessarily subjected to a nation's current development stage and/or geographical location. Nanotechnology has the asset of both, to bypass and to bridge yesterday's missing technological link between the developed and the developing world (Bürgi and Pradeep 2006).

Technological and social changes may not occur contemporaneously since the social system requires its own time to respond to alteration and to find its new equilibrium. Social change is as dynamic and complex as social systems are, and with regard to social evolution, it is both the essential ingredient and the driving force. Progressive technological and social changes do not necessarily eradicate previous forms, and historically, science and technology have been used by all kinds of societies irrespective of their stage of development and belief (Bürgi and Pradeep 2006).

The technological, social, and cultural evolution process depends on a large number of factors, and the changes modify the way the society exists. A couple of the major factors that generate this social progress are the scientific knowledge applied and the developments associated with it.

6.3   Social Acceptance of Nanotechnology

The term "social acceptance of nanotechnology" could be defined by the tolerance for the impact that nanotechnology can have on society. A nanotechnology should be defined as unacceptable if it is in contradiction with the values, interests of society, and if it can harm society or its development in the current time or in the future.

One case of use of nanotechnology is related to food with nanotechnology packaging. When you buy food that contain nanoscale additives and if the pack does not contain explicit information that these additives have benefits (Siegrist et al. 2007), the customer could opt for other products, because the price is lower or because they are habituated to buy always the same. So the acceptance could depend on the price, the benefits, and the information on the pack.

Functional food is a creation on nanotechnology that promises consumer improvements in targeted physiological functions, and the benefits cannot be directly experienced by the consumer, and the consumer must believe that they will get these benefits. This makes trust crucial for the acceptance of functional foods, because consumers must rely on the health claims provided by producers (Verbeke 2006) (Siegrist 2000).

The academia, government, and industry need to be extended to the civil society, and the public and communications media need to be involved from the beginning, since news and information regarding nanoscale science and technology will shape the public perception and determine, to a large extent, people's knowledge, attitude, and behavior toward this new technology. Consumer acceptance is the key when it comes to commercially developed nanotechnology products because ultimately it is the end users who will influence the trajectory of nanotechnology (Bürgi and Pradeep 2006).

The involvement of all sectors of the society is the key to social acceptance of nanotechnology and nanoscience because knowledge and discussion guide the acceptance. When you know that it involves a new product with new technology, it is easier to trust it, and involvement from the beginning facilitates the process of trust.

The convergence of the newly emerging technologies of the 21st century has the potential to revolutionize social and economic development and may offer innovative and viable solutions for the most pressing problems of the world community and its habitat. It will provide policymakers with better tools to take responsible choices. Nanoscience and its deriving technologies have the potential to improve the state of the developing world, if the applications are designed and tailored to best fit the needs of the people (Bürgi and Pradeep 2006).

It is imperative that genuine risks be dealt with in an expeditious, open, and honest manner. Negative public attitudes toward nanotechnology could impede research and development, leaving the benefits of nanotechnology unrealized and the economic potential untapped, or worse, leaving the development of nanotechnology to countries and researchers who are not constrained by regulations and ethical norms held by most scientists worldwide. Research on how to achieve an informed population will be important for establishing best practices for educating, communicating, and engaging diverse publics about nanotechnology. We need to develop survey data about understanding attitudes, information about audience response to various media products, and effective training methods to prepare scientists and engineers to engage in public dialog about nanotechnology (Roco and Bainbridge 2005).

Better understanding of the potential benefits and hazards of nanoscale science and technology is essential. The potential societal implications of the scientific and technological innovation process and the diffusion of the future applications of the discoveries at the frontiers of science could be, in most cases, weakly understood and lack better exploration. The characteristics of the nanoscale permeate a dimension of human life that is hard to see with the naked eye, and their impact is hardest to perceive.

As mentioned previously, the knowledge and participation of society in the discussion of the creation of new products based on nanoscience are primordial for their acceptance and trust. It is primordial to involve all sectors of the society from the beginning, and the scientific community must be involved in the prediction of the future impact.

Consumers have emphasized the importance of ethical concerns as a determinant, or otherwise, of acceptance of specific products, but expressed less concern regarding the potential physical contact with the product when compared to what had been predicted by experts. Similarly, consumers were less concerned about food-related applications of nanotechnology than expert predictions of consumer concern (Frewer et al. 2014).

The social perception that any negative and unintended biological effects can be irreversible when a nanotechnology is launched into the environment can cause fear and discontent. Because of this, all new nanotechnologies must be well evaluated, discussed, and explained to its best for understanding the consequences of its use, such as the use of genetically modified foods.

Some recently implemented food technologies, such as high-pressure processing or other cold food preservation technologies, have been accepted by both society in general and consumers, with little societal discussion of their merits or demerits. While several authors have conducted comparative reviews of research focused on consumer perceptions of, and attitudes toward, technologies applied to agrifood production, including nanotechnology, they have failed either to consider agrifood nanotechnology in detail (being focused on gene technologies applied to food production, where there are more data available), or have discussed generic attitudes toward food technologies rather than nanotechnology specifically, or have failed to consider the factors underpinning the lack of current societal discourse regarding agrifood nanotechnology relative to other, earlier, controversial food technologies (Frewer et al. 2014).

6.4   Identifying Ethical Issues

The ethical issues are the most uncertain because there is no agreement. It could be defined at first, but the most important thing is trying to define some norms to follow. The technology must have an ethical reflection for the impact on the life and environment. This impact implies in lives, animals, and vegetables, and must be well planned from research to production to market and the long-term uses that will affect the social structure and environment.

A technology that unquestionably improves the conditions of life of individuals could, at the same time, increase inequality among the general population. Because for various reasons, the benefits are not justly distributed. For instance, a nanobiotechnology-based medical treatment could be so expensive that only the economic elite can afford it; or the beneficial use of a nanotechnology-based device may require considerable knowledge skills so that in practice only the educational elite can benefit from it (Schummer 2006).

6.4.1   General Ethical Concepts

Establishing ethical boundaries together with the ethical situations involved with nanotechnology is a task that needs to be defined on a case-by-case basis. Each new product or technology has its own ethical implications. It may not be possible to just define a list of situations that should be checked and think that they will be sufficient.

It is possible to find a starting list, which should be verified but is not safe to rely only on this list. This task is complex and requires an investment of time at least adequate so that you can correct all possible situations and then mitigate or solve them and only then continue with the development process.

Biological materials, like cells, have different ethical issues, and it is more critical with the stem cells and ova fertilized or not, because they are cells that can generate life. Some cultures are against this kind of research, which will be discussed later. What will happen in the human body with a modified cell must be planned, and if it is the right use of nanotechnology, it could be applied to the cell for body physiologic modifications.

Studies on animals are very restricted now, and some institutions have a specific department to treat ethical issues related to the use of animals. It is no longer acceptable to use animals in studies, with nanotechnology, nanoscience or other, without a good reason and well-planned way to conduce the research. All the tests that might be performed without uses of animals or humans must be done and it is not acceptable to not do it because it is more expensive, because at the end all lives matter.

6.4.2 Ethical Principles

Ethical and moral considerations have been shown to influence public acceptance of nanotechnology. The ethical basis for future consideration of nanotechnology has some principles, namely, beneficence, non-malfeasance, justice, and autonomy, which are introduced as follows and must be counted for futures discussion.

- Beneficence: Any identifiable benefits associated with the technology application.
- None Malfeasance: The requirement to do no harm or at the very least minimize harms.
- Justice or fairness: Distribution of risk and benefit such that benefits do not accrue to one stakeholder, while another bears the bulk of the risk.

- Autonomy: End users, consumers, or other stakeholders are provided with sufficient information and freedom to enable them to decide whether or not they wish to adopt or make use of nanotechnology applications in the food chain (Frewer et al. 2014).

6.4.3 Ethical Issues of Modified Foods

The experience with genetically modified foods has been prominent in motivating science, industry, and regulatory bodies to address the social and ethical dimensions of nanotechnology. The overall objective is to gain the general public's acceptance of nanotechnology in order not to provoke a consumer boycott as it happened with genetically modified foods (Ebbesen 2008).

It is stated implicitly in reports on nanotechnology research and development that this acceptance depends on the public's confidence in the technology and that the confidence is created on the basis of information, education, openness, and debate about scientific and technological developments. Hence, it is assumed that informing and educating the public will create trust, which will consequently lead to an acceptance of nanotechnology. Thus, the humanities and social sciences are seen as tools to achieve public acceptance (Ebbesen 2008).

If novel agrifood technologies are perceived by consumers to act against their existing preferences (for example, through negative impacts on the environment, increased globalization of the food supply, or compromised animal welfare standards), or if consumers perceive that they have been unknowingly exposed to risky or unethical food risks associated with innovations in agricultural production, then acceptance of products may be problematic. The extent to which people perceive these foods to be unnatural and have ethical concerns about technology as "tampering with nature" is associated with higher risk perceptions and lower perceptions of benefit (Frewer et al. 2014).

6.4.4 Medical Ethical Issues

The issues regarding medical usage are always hard to discuss. Actually, the access to better treatments is available only for a small portion of the population. Since the research on new technology takes a lot of money, new treatments have a high price. Nanoscience and nanotechnology, which could repair damaged organs or the DNA in the future, to prevent genetic diseases, will cost a lot of money and will not be for everyone, unfortunately.

The situation that this issue put on the table is just about the right for a life, with quality and equality, but since when the medicine research began this is a constant, and it is not a guilt of the physician or other health professionals, is just because the money rules the destiny of all of us and the capitalist system needs to make money what makes the access to a good treatment expensive and not available for all.

6.4.5   Usability Ethical Issue

In principle, complex things, done in a more complex way, with more complex functionalities, tend to have a more complex form of use, which leads us to think about the question of the usability of nanotechnologies.

To exemplify the concept, before applying to the nanotechnology, if we take smartphones as a use case, while a large part of the population uses smartphones in a simple way, another large part of the population does not use all the capabilities of the device because they do not know how to operate it properly. And yet another large part of the world's population still uses conventional cell phones simply because they have not been able to adapt to the new technology, and thus prefer to use mechanical buttons than to use a touch screen.

With the new nanotechnologies, applied to the telecommunications and the computation, will be available new ways to make calls and a new generation of computers, but, including the price that will not be accessible for all in a short term, there is the usability issue, when not all people will be able to use it, and a question surge "Is it fair to create technology that will improve the life quality of not everyone?"

In medicine will surge new treatments based on the uses of new technologies of medical devices and new pharmacology chemical, but this new technology could not be simple to use, and could exclude patients that are not able to use it, or because their physicians are not able to use it.

In the develop of the nanotechnology we must have the care to create a thing simple to use, to evolve as many people as possible to make it accessible and usable for a major number of the population.

6.4.6   Implanted Nanochip Ethical Issues

Such a practice can transform the way individuals interact with the world around them, and it may make certain infrastructures obsolete while also creating needs for new ones. While the individual uptake of chip implants was limited to a few cases, it can be hypothesized that a more structured and organized use of these chips in the workplace can lead to a surge in chip-related services and uptake across different areas of society (Gauttier 2019).

The discussion arises about whether it is ethically acceptable the use of nanochips implants under the skin or on it, this can possibly enable the complete scan of the individual during its use, and it is very important to emphasize that the privacy of the use of the individual's data can be violated, in the other hand it turn possible things like, pandemic control, criminal tracking and other situations like prove of presence in certain place for law issues. Other situation about this is the presence detection, like a person could not open a door to enter in a restrict area, because the door knows who you are, maybe this situation could be less insecure than someone having stolen your ID card. We should have a better understanding of a multidimensional ethics to study this situation and determine the limitation of the data collection that must be used.

6.5   Addressing the Issues to the Geographical, Economic, Psycho-Social, Affective, Cognitive, Technical Administrative, and Political Parameters

Understanding the geographical, economic, psycho-social, affective, cognitive, technical administrative, and political parameters, which determine people's attitudes to and acceptance of emerging technologies and their applications, is now recognized by stakeholders in academia, industry, and policy communities as being an important determinant of their successful implementation and commercialization, because without a large investigation of these factors is like throwing the money invested on the development of nanotechnology to a luck spin.

The humanities and social sciences have a critical function asking fundamental questions and informing the public about these reflections. This may lead to skepticism; however, the motivation for addressing the social and ethical dimensions of nanotechnology should not be public acceptance but informed judgment (Ebbesen 2008).

Proposing chip implants to employees raises questions about the role that technology is meant to have in society. Technology has always been an integral part of a workplace. Technology used to take the form of an external artefact, and most of the time it would be activated by the expressed willingness of the user, who had at least a certain level of control on the technology (e.g., the ability to operate a machine functioning in the background). Moreover, the technological artefacts provided were used by individuals only for a given period during the day. Embodied technologies seem to be in opposition to the technologies described above: They are a priori characterized by a few features, namely their continuous presence and potential use, the lack of control of the user, and forms of use that do not require express intentionality. They can also be open to hacking, that is, others may modify the functionalities of the chips or to access information through them. Most importantly, they could also eventually become an intrusion into the bodily functions of the individual (Gauttier 2019).

Today technology has become an integral part of workplace, but one that is implantable opens some issues of privacy, when it could be used to monitor and control the steps of employees, and in the long term might impact on the structure and functioning of society. One thing that could change in this situation is the fact that the employees may lose contact with their employer, when the information about them comes from an implanted nanochip and not from interaction with a person.

6.5.1 Geographical Issues

Geographical perspective of nanoscience in the globalization process is related to the economy. Some regions of the world have more economic facilities that bring some factories to them. Currently, the most popular region for manufacturing products is China; reduced rates and cheap labor make this place one of the main choices for plant creation. This example points out that geographically a given region becomes more favorable to produce certain equipment, which implies social and labor problems due to cheap labor where individuals must find a job that earns them enough for survival.

The issue of geography for nanotechnology is the assessment of a cheap workforce, but other issues could be addressed to the raw material for manufacturing nanotechnologies, and it implies to an environment exploration for the natural resources necessary for manufacturing and at the end the impact of this exploration in the environment, in the natural resources and the way of life of the population who lives around the place of exploration and these issues must be well planned.

6.5.2 Economic Issues

The perception of the ethical implications of nanotechnology also depends on the economic situation of a country. If nanotechnology is considered as enabling "the next industrial revolution," i.e., as providing a unique opportunity for huge economic improvement, no country wants to lag behind. Thus, the economic promise puts enormous pressure on suppressing or at least outweighing ethical issues, both in developing and developed countries. However, there are some important differences between these two situations (Schummer 2006).

As in the geographical issues, the economic transformation of a region must be well planned. Some exploration may affect the way of economy use. Some activities could disappear, leaving a hole in the life of a small population. On the other hand, a well-planned transformation could improve the life condition of this population.

6.5.3 Psycho-Social, Affective, and Cognitive Issues

The geographic and economic transformations create a change in the way of life of some population. These transformations lead to a change in lifestyle, which leads to the extinction of certain habits and crafts and generates uncertainty and insecurity.

Emotional responses to changes might lead to some pathologies, and the most important reflection at this point is the consequences of that changes, not about what triggers which pathology, so in this way these changes might modify the way how people use to interact each other or cease this interaction due to psycho-social changes, implying in affection and cognition.

In its brief history, psychology has undergone wrenching paradigm shifts. In these transformations, the theorists and their followers think, argue, and act agentically, but their theories about how other people function grant them little, if any, agentic capabilities. It is ironic that the science of human functioning should strip people of the very capabilities that make them unique in their power to shape their environment and their own destiny (Bandura 1999).

The behaviorists gave us the input–output model linked by an obscure black box. In this view, human behavior is conditioned and regulated by environmental stimuli. This line of theorizing was eventually put out of vogue by the advent of the computer, which filled the black box with a lot of self-regulatory capabilities created by inventive thinkers. One brand of behaviorism survived with an even more stringent orthodoxy in the form of the operant model of human behavior. Operant conditioners not only stripped human beings of any agentic capabilities but also imposed strict methodological prohibitions that even natural scientists reject (Bandura 1999).

This may lead to a reflection of these issues that could be addressed in nanotechnology, during the process of resources exploration and fabrication. But one last reflection is that the commercialization of new products, in this case nanotechnology products, could cause a change and might cause expected or unexpected psycho-social, affective, and cognitive side effects.

6.5.4 Technical Administrative Issues

It is a false premise to assert that workers have free choice in terms of which work and working conditions to accept. Although some component of self-determination is present, economic and social conditions exert the greatest influences on workers' selection of work, level of risk tolerated, and ability to participate in risk management (Schulte and Salamanca-Buentello 2007).

Workers generally cannot universally refuse work they consider hazardous and still keep their jobs. Conformance with the principle of autonomy depends on the extent to which workers have input into risk management at their work sites and the degree to which they are at risk after controls have been implemented. Justice is also related to worker decision-making (Schulte and Salamanca-Buentello 2007).

A technical administrative issue is the extent to which workers are exposed to greater risks than the general public and the need to manage this situation. On one hand, work needs to be done; on the other hand, there is health risk. So some incentives such as wages or hazardous duty pay for additional risk from exposure to nanoparticles should be created to remedy the situation. But the work continues to be at risk; there exists a tolerance level to a protection equipment to reduce this risk, and this is another ethical issue that sometimes passes by for some companies because the work needs to be done.

6.5.5 Governance and Political Parameters

Countries differ in their general political cultures and systems. Provided that citizens trust their general political system, any form of technology governance that does not fit the general political system may cause mistrust. Thus, citizens in a strongly democratic system would mistrust the autocratic model of technology governance, and vice versa. Moreover, societies differ in the degree of desired political regulation. Some countries prefer less political control and planning, relying more on free market control. For such countries, both autocratic and democratic models of technology governance would be foreign, whereas the information-plus-debate model that educates informed consumers would appear more suitable. Other countries trust more in the efficacy of political control and advance planning, for which the autocratic or democratic models of technology governance would be more suitable than the information-plus-debate model (Schummer 2006).

This means that the political interests that affect the education system and others will interpose in the culture of a population and affect the way the population, in general, will think about some information or technology. Nanoscience could be perceived in different ways according to the political system. Moreover, it is the responsibility of the political system to manage the way in which a specific technology operates in a country or continent.

Here are some responsibilities of the government with respect to nanotechnology. These items must be monitored by the third sector, and this should not neglect them to the detriment of financial advantages.

- Identification of benefits associated with the (specific) application of nanotechnology.

- Identification of the risks associated with the (specific) application of nanotechnology.

- Differential accruement of risk and benefit associated with a (specific application of) nanotechnology to different stakeholders or groups in the population.

- End users, consumers, or other stakeholders can choose whether to adopt, be exposed to, or utilize (a specific application of) nanotechnology.

6.6   Ethical, Religious, and Cultural Acceptability

Studying past examples of innovations may help to create analogies and frameworks for understanding and anticipating some issues of ethical acceptability when looking through the religious and cultural prism. Since patterns of cultural and religious responses might influence the outcomes of new technology acceptance, it is important to have a thought in the past examples of other technologies (not nano) to anticipate the future issues.

Cultures that value justice over utility will certainly raise ethical concerns about the injustice induced by the new technology. Others that put a lower value on justice, or have a different concept of justice, will embrace the technology without much hesitation. Cultures with a still lower evaluation of justice would perhaps accept the technology even if it posed unequally distributed risks, such that it benefited a fraction of society and harmed another fraction, so long as the benefits overall outweighed the harms (Schummer 2006).

Research into consumer perceptions and attitudes has focused on general applications of nanotechnology. Perceptions of risk and benefit associated with different applications of nanotechnology shape consumer attitudes and acceptance, together with ethical concerns related to environmental impact or animal welfare. Attitudes are currently moderately positive across all areas of application. The occurrence of a negative or positive incident in the agri-food sector may crystallize consumer views regarding acceptance or rejection of nanotechnology products (Frewer et al. 2014).

There are many different cultures and religions. Some rituals and practices differ from one another, and these differences may affect the acceptability of a new technology, and nanotechnology. The fact of the nanotechnology is in a nano size, and hard to see it, might has benefits, or because this, it might causes discomfort or distrust, generate fear and misconceptions about its real purpose.

As mentioned earlier about the nanochips under the skin, the use of data may create ethical embarrassment and lead to refusal of their use. It is mandatory to provide well explanation of their use and purpose as well as clear objectives so that their users have knowledge of what they are acquiring, how it will work, and what are the implications of its use, be it in terms of data use, privacy, and mainly the risks involved.

As with many other political and scientific issues, citizens rely on cognitive shortcuts or heuristics to make sense of issues for which they have low levels of knowledge. These heuristics can include predisposing factors, such as ideological beliefs or value systems, and also short-term frames of reference provided by the media or other sources of information (Scheufele et al. 2009).

Recent research suggests that "religious filters" are an important heuristic for scientific issues in general, and nanotechnology in particular. A religious filter is more than a simple correlation between religiosity and attitudes toward science: It refers to a link between benefit perceptions and attitudes that varies depending on respondents' levels of religiosity (Scheufele et al. 2009).

In surveys, seeing the benefits of nanotechnology is consistently linked to more positive attitudes about nanotechnology among less religious respondents, with this effect being significantly weaker for more religious respondents. The results show that respondents in the United States were significantly less likely to agree that nanotechnology is morally acceptable than respondents in many European countries. These moral views correlated directly with aggregate levels of religiosity in each country, even after controlling for national research productivity and measures of science performance for high-school students (Scheufele et al. 2009).

Societies also differ greatly in their normative ideas about human identity and integrity, putting different weight on different aspects of human existence, and accordingly their perceptions of ethical issues of nanotechnology differ (Schummer 2006).

The impacts of ethical religiosity and culture on the acceptability of nanoscience and nanotechnology leads using a relatively comparison of the impacts of other past knowledge and technology in an attempt to decipher how the future of nanotechnology and nanoscience might be, there is no practical or manual guide to be followed exactly, but there is a wide variety of information from the past that might guide this process.

References

Bandura, Albert. 1999. "Social Cognitive Theory: An Agentic Perspective." *Asian Journal of Social Psychology* **2** (1): 21–41.

Bürgi, Birgit R, and T Pradeep. 2006. "Societal Implications of Nanoscience and Nanotechnology in Developing Countries." *CURRENT SCIENCE-BANGALORE-* **90** (5): 645.

Ebbesen, Mette. 2008. "The Role of the Humanities and Social Sciences in Nanotechnology Research and Development." *NanoEthics* **2** (1): 1–13.

Frewer, Lynn J, N Gupta, S George, ARH Fischer, El L Giles, and David Coles. 2014. "Consumer Attitudes Towards Nanotechnologies Applied to Food Production." *Trends in Food Science & Technology* **40** (2): 211–25.

Gauttier, Stéphanie. 2019. "I've Got You Under My Skin, the Role of Ethical Consideration in the (Non-) Acceptance of Insideables in the Workplace." *Technology in Society* **56**: 93–108.

Heberle, Sandra Magali, Gustavo Marques da Costa, Nelson Barros, and Michele SG Rosa. 2018. "The Effects of Atmospheric Pollution in Respiratory Health." In *Handbook of Environmental Materials Management*, 1–16. Springer International Publishing Cham.

Hussain, Chaudhery Mustansar. 2018. *Handbook of Nanomaterials for Industrial Applications*. Elsevier.

———. 2020a. *Handbook of Functionalized Nanomaterials for Industrial Applications*. Elsevier.

———. 2020b. *Handbook of Manufacturing Applications of Nanomaterials*. Elsevier.

———. 2020c. *The ELSI Handbook of Nanotechnology: Risk, Safety, ELSI and Commercialization*. John Wiley & Sons.

Panis, L Int, Carolien Beckx, Steven Broekx, Ina De Vlieger, Liesbeth Schrooten, Bart Degraeuwe, and Luc Pelkmans. 2011. "PM, NOx and Co2 Emission Reductions from Speed Management Policies in Europe." *Transport Policy* **18** (1): 32–37.

Riley, Janet. 2001. "Indicator Quality for Assessment of Impact of Multidisciplinary Systems." *Agriculture, Ecosystems & Environment* **87** (2): 121–28.

Roco, Mihail C, and William Sims Bainbridge. 2005. "Societal Implications of Nanoscience and Nanotechnology: Maximizing Human Benefit." *Journal of Nanoparticle Research* **7** (1): 1–13.

Scheufele, Dietram A, Elizabeth A Corley, Tsung-jen Shih, Kajsa E Dalrymple, and Shirley S Ho. 2009. "Religious Beliefs and Public Attitudes Toward Nanotechnology in Europe and the United States." *Nature Nanotechnology* **4** (2): 91–94.

Schulte, Paul A, and Fabio Salamanca-Buentello. 2007. "Ethical and Scientific Issues of Nanotechnology in the Workplace." *Environmental Health Perspectives* **115** (1): 5–12.

Schummer, Joachim. 2006. "Cultural Diversity in Nanotechnology Ethics." *Interdisciplinary Science Reviews* **31** (3): 217–30.

Siegrist, Michael. 2000. "The Influence of Trust and Perceptions of Risks and Benefits on the Acceptance of Gene Technology." *Risk Analysis* **20** (2): 195–204.

Siegrist, Michael, Marie-Eve Cousin, Hans Kastenholz, and Arnim Wiek. 2007. "Public Acceptance of Nanotechnology Foods and Food Packaging: The Influence of Affect and Trust." *Appetite* **49** (2): 459–66.

Verbeke, Wim. 2006. "Functional Foods: Consumer Willingness to Compromise on Taste for Health?" *Food Quality and Preference* **17** (1-2): 126–31.

Wallington, Timothy J, John L Sullivan, and Michael D Hurley. 2008. "Emissions of Co2, CO, NOx, HC, PM, HFC-134a, N2o and Ch4 from the Global Light Duty Vehicle Fleet." *Meteorologische Zeitschrift* **17** (2): 109–16.

Wood, Stephen, Alison Geldart, and R Jones. 2003. "The Social and Economic Challenges of Nanotechnology." *TATuP-Zeitschrift für Technikfolgenabschätzung in Theorie Und Praxis* **12** (3-4): 72–73.

Chapter 7

Ethical Issues in Nanotechnology-II

**Wilson Engelmann, Raquel von Hohendorff,
and Daniele Weber da Silva Leal**
*Universidade do Vale do Rio dos Sinos (UNISINOS),
95 Unisinos Avenue, São Leopoldo, RS 93022-750, Brazil.*
wengelmann@unisinos.br

Promoters of nanotechnology agree that the nanoscale has opened new opportunities for inventions that enhance existing technologies; from fields such as microcomputer, automotive industries, and defense to agriculture, health, and office automation, all sectors will be affected. Nanotechnologies have the potential to transform all sectors of human experience, making it difficult to understand the "why" and "how" of the accommodation of their own ethics. Changes in the essence of the nature of human actions, which have caused a transformation relative to generating destructive effects on the entire biosphere of the planet, have created the need for "new ethics" anchored in the Fourth Industrial Revolution. All technical action has implied a failure to distinguish the good from the bad. This ambivalence has resulted in unpredictability concerning the effects that technical human actions may generate.

Environmental, Ethical, and Economical Issues of Nanotechnology
Edited by Chaudhery Mustansar Hussain and Gustavo Marques da Costa
Copyright © 2022 Jenny Stanford Publishing Pte. Ltd.
ISBN 978-981-4877-76-3 (Hardcover), 978-1-003-26185-8 (eBook)
www.jennystanford.com

When the European Union studied emerging fields in science and technology, they perceived developing fields in nanosciences and nanotechnologies with the potential to have a positive economic, social impact. There are still knowledge gaps about the negative impact of these technologies on human health and the environment as well as on issues related to ethics and respect for fundamental rights. Since 2008, the European Commission has recommended that the Member States adopt the Code of Conduct to govern research in this field. Based on approximately seven general principles covering issues such as sustainability, precaution, inclusion, and responsibility, the Code of Conduct invites Member States to take concrete measures for the development and safe use of nanotechnologies. Among these principles is sustainability. The code states that nanoscience and nanotechnology research must be safe, ethical, and contribute to sustainable development, in agreement with Sustainable Development Goals (SDGs). Goal 4 of the SDGs, which deals with the necessary support for responsible development, must be respected. The National Nanotechnology Initiative 2020 Report refers to concerns about the ethical, legal, and social implications of nanotechnology. Where ethics in research are concerned, one should keep in mind two aspects: (1) research integrity: the prevention of unacceptable research practices, and (2) science and society: the ethical acceptability of scientific and technological developments. Ethics should not be perceived as a restriction on research and innovation, but as a way to guarantee high-quality results. We propose a study of the fundamentals of "nanoethics," a combination consisting of the benefit/risk approach, concerns regarding good "governance" of innovations, and a context in which science may or may not be strongly contested. Here, based on John Finnis' "coherent life plan," we use key points to mediate the individual and the collective, the social acceptance of risk, and the intent to design a model, in the hope of equating the costs and the benefits of nanotechnologies.

7.1 Introduction

The 21st century has marked the consolidation of the so-called Fourth Industrial Revolution [26]. Within the panorama of the technologies and innovations that have arisen during this revolution, we find nanotechnologies, defined as a set of research, development, and innovation actions obtained through the special properties of matter organized from structures of nanometric dimensions [16]. A nanometer (1 nm) equals a billionth of a meter, which, according to an example from the manual published by the Brazilian Agency for Industrial Development (Cartilha da ABDI), is about 100,000 times smaller than the diameter of a strand of hair, 30,000 times smaller than one of the strands of a spider web or 700 times smaller than a red blood cell [1].

This scenario of great technological development—in which nanotechnologies offer the promise of innumerable benefits in the most diverse areas—has emanated without the necessary scientific knowledge of its impacts and effects. Consequently, risks concerning possible irreversible damage to the environment and human life are now ensconced in the same context. Such potential risks regarding future damage require the establishment of boundaries for ethical conduct, a task which dates back to the origins of bioethics. In recent years, there has been a progressive ethical awareness of the various challenges produced by scientific advances as well as by economic and technical progress. This brings us to the idea that not all scientific discoveries and not all technological advantages generate purely beneficial effects. Current ecological concerns have led to the emergence of bioethics that strongly intervenes in life sciences and health policies with the aim of highlighting ethical issues [17].

It is precisely within this universe that nanotechnologies have been operating and increasingly producing nanomaterials, despite the unanswered questions about their future impact on human life (such as cases of irreversible damage). Even in the face of these unknowns, nanoscale technology has been widely used. As such, only with proper reflection, in addition to the application of bioethics, will it be possible to carry out the production of nanomaterials more consciously while respecting fundamental rights, especially the founding one: human dignity. As Vicente de Paulo Barretto has taught us, it is a clear case in which the classic legal model is insufficient but in which the law, as the most important source of the Brazilian Legal System, presents some answers for facing social problems. In addition to these possible gaps, one must not forget the need to build a legal order that seeks to legalize the new demands of a techno-scientific society. Contemporary reality has done away with the pretensions of utilitarian rationality of law, as has been clearly shown by the more radical positivist model. Accordingly, it is reasonable to question to what extent the idea of responsibility can legitimate the regulation of margins of freedom [4]. It is in this scenario that nanotechnologies are inserted and must be unveiled, through bioethical principles and reflections, to achieve development accompanied by guarantees of protection of human dignity.

Bibliographic research will be used to review publications in books and scientific articles as well as on official internet sites. This methodological resource will be allied with content analysis, along the lines of the analysis presented by Laurence Bardin in the book she named after the same research tool [3]. As a starting point, we must address previous knowledge about nanotechnologies and risk, to provide the background for the need to impose limits through bioethics and, therefore, safeguard human dignity.

The problem that this chapter aims to address may be thus defined: under what conditions can an ethico-legal perspective use a variety of bases and principles to enable the development of nanotechnology in favor of the protection of human dignity? Furthermore, can it standardize the management of possible risks throughout the life cycle of engineered nanomaterial?

The provisional hypothesis was structured from a transdisciplinary perspective and comprehends the following structural elements: The European Union, upon studying emerging fields of science and technology, identified developing fields in nanoscience and nanotechnology with the potential to exert a positive economic, social, and environmental impact. This investigation may guide techno-scientific development in Brazil. There are still knowledge gaps about the negative impact of these technologies on human health and the environment, as well as on issues related to ethics and respect for fundamental rights. Since 2008, the European Commission has recommended that the Member States adopt the Code of Conduct to govern research in this field. Based on approximately seven general principles covering issues such as sustainability, precaution, inclusion, and responsibility, the Code of Conduct invites Member States to take concrete measures for the development and safe use of nanotechnologies. Based on this ethico-legal model, normative alternatives for the Brazilian legal system will be proposed, using the methodological framework designed by John Finnis, through which he presents the so-called "coherent life plan." Among these principles is sustainability. The code states that nanoscience and nanotechnology research must be safe, ethical, and contribute to sustainable development, in accordance with Sustainable Development Goals (SDGs). These activities must comply with SDG Objective 4, which deals with the necessary support for responsible development. The National Nanotechnology Initiative 2020 Report (USA) refers to concerns about the ethical, legal, and social implications (ELSI) of nanotechnology. Where research ethics are concerned, one should bear in mind two aspects: (1) research integrity: the prevention of unacceptable research practices, and (2) science and society: the ethical acceptability of scientific and technological developments. Consequently, from the components of ELSI, we will seek to structure normative proposals for nanotechnologies. In order to proceed with the testing of this hypothesis, the following chapters have been composed.

7.2 Insertion of Nanotechnologies into Human Life and the Risks: The Emergence of an Ethico-Legal Category

New technologies, developed by the most diverse fields of human knowledge, challenge our ability to understand the world in which every one of us lives. The advances generated by technological fields (hard science fields) must be reinforced by the epistemological tenets forged by the human sciences. Therefore, it seems that, more and more, the dichotomy between nature and culture has lost its vigor and raison d'etre. In the universe of nanotechnologies, nature and culture have grown progressively closer, under their reciprocal influences. Not long ago, one would speak about microscopic discoveries; today one speaks about discoveries at the nanoscale. This, of course, causes curiosity and broadens challenges and clearly shows how culture (here represented by science) filters down into natural structures at progressively smaller scales, with the aim of extracting the largest quantities of benefits [12].

As Engelmann [13] has stated about nanotechnologies, the "so-called future has already begun," as they are in a constant and increasing process of induction, despite the fact that most of the risks that this set of novelties may produce are still unknown. However, the best option is to be neither against nor promote postponement:

> In public engagement, we catch a glimpse of an appropriate path, in the sense of democratically accompanying the assessment of the social, environmental, and legal impacts of the new nanotechnologies. Monitoring and compliance with the precautionary principle will be necessary for the mitigation of the negative effects that inevitably accompany any novelty.

Here, there is ambivalence among the various technologies and products generated at the nanoscale. While possibilities and positive effects have been assessed as considerable, the probability of risks is very significant to the ecosystem in general. Current research developed so far has contributed to this concern and has generated a closer relationship between countries that seek to produce nanotechnology while preserving the environment [13]. This is a great challenge that bridges ethical and legal concerns. Despite this scenario, the number of products that reach the consumer market is growing constantly. In a survey carried out by June 5, 2020, in the Nanotechnology Products Database (NPD), the following results were found (Table 7.1):

Table 7.1 Available data on nanoproducts

Products	9,015	Products from various sectors, as will be seen below, using some kind of nanoparticle.
Manufacturers	2,453	These products are distributed in the following segments: agriculture, automobile, construction, cosmetics, food and packaging, household appliances, sports and fitness, textiles, electronics, oil, medical equipment, medicines, among others.
Countries where these companies are located	62	

Since nanoscale production manipulates different characteristics of each substance, in this diverse set of segments, different results have been obtained that were not previously present in conventional equivalents. Therefore, this status demonstrates that nanotechnologies are a form of human intervention in the system of nature. Thus, the manipulation of atoms and molecules provides new challenges for humans, including the creation of unprecedented rights and obligations, since the effects of "Nanos" on human life are unknown. This context is characterized by the unknown potential of nanotechnology, an unknown that generates uncertainty, which then flows into risks [9, 14, 28].

Therefore, according to the teachings of Ulrich Beck [5], the insertion of nanotechnologies into Risk Society is undeniable: in the modern world, the difference between the language of quantifiable risks, which one operates, and the world of unquantifiable uncertainty, which one creates, can expand considerably, following the pace of technological development. Therefore, decisions made in the past with regard to nuclear energy and current ones, such as those concerning engineering and the exploration of genetic engineering, nanotechnology, information technology, and so on, are triggers of unpredictable, uncontrollable, and even incommunicable consequences that threaten life on our planet. According to Beck [5]:

> Nel mondo moderno, il divario tra lingua dei rischi quantificabili, in base ai quali pensiamo e operiamo, e il mondo dell'incertezza non quantificabile, che abbiamo creato noi stessi, si amplia sempre piú, seguendo il ritmo dello sviluppo tecnologico. Le decisioni che abbiamo preso in passato in materia di energia nucleare e quelle attuali in merito allo sfruttamento dell'ingegneria e della manipolazione genetica, della nanotecnologia, dell'informatica e cosí via scatenano conseguenze imprevedibili, incontrollabili e addirittura incomunicabili, che minacciano la vita sul nostro pianeta.

Ulrich Beck addresses the risks brought about by nanotechnologies as he discusses his perspective. Moreover, he finds that nanotechnologies generate a true metamorphosis of the world: When we talk about "change in society," a future is projected that presents itself as a characteristic of modernity, that is, "permanent transformation," while the basic concepts and certainties that sustain them remain constant [6]. However, the convergence of technologies at the heart of the Fourth Industrial Revolution [27] has shown us something different; something which Ulrich Beck calls a "metamorphosis of the world": where the certainties of modern society are destabilized. "Metamorphosis in this sense simply means that what was unthinkable yesterday is real and possible today" [6]. He points out that nanotechnologies can help to structure the metamorphosis of the world: The development of equipment and methodologies to access the manometric scale opens up the possibility of building or reconstructing whatever human beings want, including life itself. Therefore, having an architecture of ethical and legal elements is imperative to ensure some fundamental principles, such as caring for life and preserving the environment.

Since there is no specific regulation on such new technology, people and the environment remain at the mercy of unrestrained development, which can cause irreversible damage. Therefore, it is crucial that such scientific progress is observed (i.e., limited) through the premises of bioethics, by performing an ethical reflection on its principles and premises. In view of an understanding of innovations based on this, the advance of nanotechnologies will be allowed to proceed with caution and, above all, with respect for human dignity.

In order to fulfill these ethico-legal perspectives, it will become necessary to value and evaluate the ethical, legal, and social impacts (ELSI) in research, development, production, commercialization, and other stages of the life cycle of nanomaterials. A methodological path has opened for the aggregation of various areas of knowledge, based on a methodology of transdisciplinarity with a focus on sustainability. This permeates, approximates, and connects the scientific investigations carried out by the areas of knowledge involved with the nanoscale, including applied social sciences, a field to which law belongs.

Although this is an important indicator, much remains to be done. Even though transdisciplinarity has become increasingly visible as a general approach to address the shortcomings of prevalent methodologies and modes of organization in scientific research, the transdisciplinarity discourse has not yet managed to develop a clear, unambiguous course. In particular, little acknowledgment has been given to the importance of a reflective questioning of values, background assumptions, and normative orientations of various approaches to sustainability in transdisciplinary research. Indeed, despite having challenged the influential conception of science as a value-neutral inquiry under the exclusive responsibility of highly trained and specialized experts, the prevalent sustainability discourse continues to build scientific reliability and social legitimacy as distinct requirements that have to be pursued in parallel and traded off against each other [24]. Hence, despite many advances, truly transdisciplinary research is still far from producing results oriented by human and environmental sustainability and this, ultimately, should be a major goal of any legitimate scientific research.

These new perspectives on research, development, and innovation are fundamental for dealing with the risks of future damage in an ethico-legal context, damages that may be generated by the unknowns regarding nanomaterials. Ulrich Beck states that "metamorphosis is linked to the idea of ignorance, which leads to a paradox: on the one hand, it emphasizes the limitations of knowledge, in particular on nanotechnology, bioengineering and other types of emerging technology which contain not only recognized risks but also risks that we cannot yet know, thus providing a window of fundamental limitations for society's ability to perceive and govern risks" [6].

In order to understand the "risk" category, a distinction between risk and danger must be made while highlighting the importance of the uncertainty regarding future damage. The potential is a result of decisions, and so the term "risk" is used and, more specifically, "decision risk." Or it is well understood that potential damage is caused externally; that is, it is attributed to the environment, and in this case, "danger" is mentioned. In this manner, the risk is associated with the decision, expectation, and probability of things that will happen in the future. It is communication focused on the future [18]. These are the characteristics of the "risk" category, which must be understood and standardized: It is no longer a take on a social fact that occurred in the past. What you may have is future damage. Certainty and predictability—structural elements of law, from its positivist roots—are replaced by probability and expectation. To deal with these changes, regulatory possibilities emerge upon looking at the philosophy of natural law presented by John Finnis.

7.3 Shifting from "The Coherent Life Plan" (Finnis) Model to the "Coherent Ethico-Legal Plan" Model to Regulate Challenges Brought about by Nanomaterials

John Finnis, in his book *Natural Law and Natural Rights*, considers the possibility of free choice as an important characteristic of humankind, especially in the execution of choices through means considered reasonable. Another aspect that draws attention is temporality, the extent to which choices assess the era of behavior that executes them. These are two characteristics suitable for the architecture of a normative plan for nanotechnologies. To develop his idea about the natural law, Finnis [15] uses the formula of the basic forms of human goods—life, knowledge, games, the aesthetic experience, sociability (friendship), practical reasonableness, and religion—and also the demands of practical reasonableness—a coherent life plan; no arbitrary preference for values; no arbitrary preferences for people; selflessness and commitment; the (limited) relevance of consequences: efficiency within the limits of good sense; respect for each basic value in each act; the requirements of the common good; following the dictates of one's own conscience; morality, as one product of these requirements.

A line of thought that seems to guide the proposal of Finnis [15] and that requires the respect of the jurist is put in the following terms: "in relation to law, the most important things that the theorist must know and describe are things that, according to their judgment, are important from a practical point of view of the law - therefore, the things that in practice are important to address while sorting out human affairs." This concern was present in the formulation of Finnis' proposal; namely, aspects that are considered fundamental and that the law should respect and take into consideration once they are linked to human issues. This is the point of convergence for using his proposal to design and structure ethico-normative models to manage the new challenges brought about by nanotechnology.

Finnis' "coherent life plan" places emphasis on the idea that each human being should have a set of purposes and guidelines that are arranged with a certain degree of harmony and allow the subject to implement them. It is not about imaginary or impossible plans, but effective commitments [15].

While the life of a human being must be guided by one project and grounded in personal past and present experiences, it should also face the future. That is to say, it is irrational for humankind to only think about the immediate moment and to forget to connect it with the others in our continuous journey in a social context.

These guidelines for humankind also serve structured social organizations: Here we may think about companies that operate at the nanoscale and how they can build a "coherent ethico-legal plan," bearing in mind the structuring elements designed by Finnis for the "coherent life plan." This plan, in spite of its well-thought-out coherence, can be subject to reforms or reviews through which old projects are left behind and new ones are chosen. Human goods are integrated and interact with a given life plan and can justify changes mid-execution or from the moment in which a new plan is drawn, especially to accommodate some human good that may need more attention at the time. This is also a characteristic of the "coherent ethico-legal plan" mode: It is marked by regulations that are flexible and adaptable to the evolution of knowledge about risks and other impacts brought about by accessing and manipulating the nanometric scale.

A rational life plan implies responsibility on the part of humankind, which should be remembered both at the time of stipulating the goals and projecting results of triggered actions. In this respect, a characteristic aspect emerges: Just as all people are not the same, plans chosen by each one also vary. This highlights the personality of choices, linked to a series of individual profiles, marked by the so-called life experience. This can be called a pre-understanding that is deeply influenced by the tradition in which the subject is immersed [11]. The reasonableness introduced in this context indicates that a coherent life plan is not just linked to individual conceptions. On the contrary, it must be designed with a view of the collective, of which the individual is a part. Thus, drawn from tradition and sensitivity to the profiles present in the world of life, a coherent life plan suggests a model of action guide decision-making about the possible risks brought about by the Nanotechnoscientific Revolution.

Hence, the formulation of a coherent life plan for the social group should be added to another basic requirement of practical reasonableness, one which John Finnis calls "the common good." This is to reinforce the need for coherent plans for the lives of human beings in society, where the individual is combined with the collective and the public with the private, without any preeminence among them. The "common good" refers to the factor or set of factors that present in a person's practical reasoning and would give meaning to, or a reason for, their collaboration with others. They would similarly give themselves, from the point of view of others, a reason for collaboration among group members and with that person [15]. The common good represents initiatives that aim at individual and collective personal well-being; that is, they seek to achieve and respect basic human goods: "what the reference to rights entails in this outline is simply an emphatic expression of what is implied in the term 'common good', namely, that the well-being of each and every person, in each of their basic aspects, must be considered and favored at all times by those responsible for coordinating common life" [15].

It should be noted that each of the basic human goods can be considered part of the construction of the "common good," insofar as they admit the participation of countless numbers of people, in the most varied forms of possibilities or occasions. Furthermore, there is another conception of the common good that accompanies Fnnis' theory: a set of conditions that enable members of a community to achieve reasonable goals for themselves, or to reasonably realize the value (or values) for themselves), for which they have reason to collaborate with each other (positively and/or negatively) as a community [15]. Because of this point of view, Finnis' outstanding ideas align with the format of the "coherent ethico-legal plan"— to be characterized below—so that actors associated with nanotechnologies, especially industries, can use anticipatorily precautionary attitudes to format managerial elements for eventual risks arising from research and product development conducted at the nanoscale. Furthermore, with their connection to the "common good," the "coherent ethico-legal plans" will be able to mediate any tension that may arise in public or in private. In this manner, the private will not prevail over the public or vice versa.

One of the guidelines that will be included in the "coherent ethico-legal plan" with regard to nanotechnologies can be qualified with the following: "[...] it is necessary to resist dehumanization; hold the actors involved accountable and anticipate the risks that may come" [10]. Technological advances (New Emerging Science and Technology [NEST]) can only be accepted as legitimate if the research and innovation resulting from them are responsible. To this end, human rights must substantiate the ethical presupposition for structuring the "coherent ethico-legal plan," considering the positive and/or negative effects linked to the nanoscale. It will be necessary to protect this content (which is equivalent to the movement against dehumanization and stands as its main focus. The humanity of humankind must be preserved). Moreover, it will be necessary to respect it; that is, to put it into practice (which means anticipating the risks that nanotechnologies may generate for humans and the environment) and establish remedial measures if they are neglected (which represents the responsibility of the actors involved).

As an example of NEST ethics, nanoethics reproduces general standards to some extent but also modifies them. There is a co-evolution between ethics and new technologies, although there are recurrent patterns of moral argument, continual learning, changes in repertoires, and new questions arise. The acceptance of preventive approaches today is widespread [29]. What one sees now with nanotechnology is more of a precaution: Promoters do not want deadlocks such as those that occurred with green biotechnology. As a result, they strive to communicate with the various actors involved in the life cycle of nanomaterials, characterizing one more element of the "coherent ethico-legal plan."

In addition to these elements, the aforementioned plan also incorporates the recommendations of the European Commission: Since 2008, Member States have signaled their adoption of the Code of Conduct to govern nanoscale research. Based on approximately seven general principles covering issues such as sustainability, precaution, inclusion, and responsibility, the Code of Conduct invites Member States to take concrete measures and involve universities, research institutes, and organizations in the development and safe use of nanotechnologies [7].

This Code of Conduct provides guidelines for specific audiences such as the Member States, employers, research financiers, researchers as well as for a wider public of individuals and civil society organizations involved in or interested in nanomaterial research (all stakeholders). These guidelines favor a responsible and open approach. The Code of Conduct invites all stakeholders to act responsibly and to cooperate with each other, for the purpose of sustainable economic, social, and environmental development. The general principles of the Code of Conduct are (1) meaning: research activities on nanomaterials must be understandable to the public, respect fundamental rights, and be conducted in the interests of the well-being of individuals and society in their design, implementation, dissemination, and use; (2) sustainability: research activities at the nanoscale must be safe, ethical and contribute to sustainable development, in accordance with the sustainability objectives of the community and the United Nations Millennium Development Goals (Goal 4 of the SDGs, which addresses the necessary support for responsible development). These investigations must not harm or create a biological, physical, or moral threat to people, animals, plants, or the environment, now or in the future; (3) precaution: nanoscale research activities should be conducted in accordance with the precautionary principle, anticipate possible environmental, health, and safety impacts of results, and take appropriate precautions, proportionate to the level of protection while promoting progress for the benefit of society and the environment; (4) inclusion: the governance of research activities using the nanoscale must be guided by the principles of openness to all interested parties, transparency and respect for the legitimate right of access to information. It must allow all interested parties involved or concerned with these research activities to participate in the decision-making processes; (5) excellence: research activities at the nanoscale must meet the best scientific standards, including standards that support research integrity and standards related to good laboratory practices; and (6) innovation: the governance of research activities in the field of nanomaterials should encourage maximum creativity, flexibility, and planning capacity for innovation and growth [8].

The National Nanotechnology Initiative 2020 Report [21] refers to concerns about the ethical, legal, and social implications of nanotechnology. Whenever ethics is mentioned in the research, one should keep in mind two aspects: (1) research integrity: the prevention of unacceptable research practices and (2) science and society: the ethical acceptability of scientific and technological developments. Ethics should not be perceived as a restriction on research and innovation, but as a way to guarantee high-quality results.

Along with all these elements, the "coherent ethico-legal plan" should also integrate the understanding and mitigation of potential impacts on the work environment. Here, one could refer to the recommendations of the National Institute for Occupational Safety and Health [20], from the United States. It should also be possible to assess the importance of adding the development of methodologies and standards to support nanoscale research and associated environmental, health, and safety impacts [21]. Moreover, the development and dissemination of information on technological matters [25] that can be understood even by lay people should similarly be included.

7.4 Final Considerations

There have been several publications on the so-called ethics of nanotechnologies [23], with each of the approaches starting from different theoretical assumptions. This chapter sought to find common ground between these diverse conceptions. For this reason, we proposed to design a "coherent ethico-legal plan" to deal with the issues of "nanoethics" and their close relations with legal perspectives. One could also ask whether nanotechnologies would bring new problems to be considered through ethics [2]. These are important issues that we intended to resolve based on the philosophical contributions of John Finnis, especially through his structure of a "coherent life plan," which has now become a "coherent ethico-legal plan." Increasingly, ethical issues have intersected with legal ones.

For this reason, the problem that this chapter intended to answer was stated in the introduction: "under what conditions can an ethico-legal perspective use a variety of bases and principles to enable the development of nanotechnology in favor of the protection of human dignity? Furthermore, can it standardize the management of possible risks throughout the life cycle of engineered nanomaterial?" The structuring of a legislative regulatory framework still seems to be a long way off, and it is still unclear whether there will ever be an agile and flexible legislative process to standardize the wealth and variety of possibilities produced by the nanoscale.

Thus, self-regulatory structures, which are flexible and adaptable, seem to be a good alternative, and this is the proposal of the "coherent ethico-legal plan." Here are the elements studied in this chapter, which help to regulate the matter, under the ethical guidance provided by two lighthouses: concern for the well-being of humankind and the preservation of the environment. As long as there are still no precise answers that are structured through transdisciplinary dialogues, our studied coherent plan can serve as a safe guide for the development of research, production, commercialization, and the final disposal of residues and packaging containing any type of nanoparticle.

Appendix A   About the object of the study in this article

A partial result of investigations carried out by the authors within the scope of the following research projects: (a) 02/2017 - Gaucho Researcher - PqG; Project Title: "The self-regulation of the final destination of nanotechnological waste," with financial support granted by the Fundação de Amparo à Pesquisa no Estado do Rio Grande do Sul - FAPERGS; (b) CNPq n. 12/2017 - Research Productivity Grants - PQ; project entitled: "Nanotechnologies and their applications in the environment: between risks and self-regulation"; (c) MCTIC / CNPq Nº 28/2018 - Universal/Band C, project entitled: "Observing Nanotechnologies and Human Rights from the risks in the panorama of communication between the Regulatory Environment and the Science System"; (d) "The Law System, new technologies, globalization and contemporary constitutionalism: challenges and perspectives," FAPERGS/CAPES 06/2018 - Postgraduate Internationalization Program in RS; (e) "Transdisciplinarity and Law: building legal alternatives or the challenges brought about by new technologies," with financial support granted by the Fundação de Amparo à Pesquisa no Estado do Rio Grande do Sul- FAPERGS, 04/2019, Aid for New Ph.D. Students.

Also, this work is also related to research carried out and presented at the Gracious Consortium meeting, "Grouping, read-across, characterization and classification framework for regulatory risk assessment of manufactured nanomaterials and safer design of nano-enabled products," with financial support from the European Union's Horizon 2020 research and innovation program, under Grant Agreement n. 760840, available at: www.h2020gracious.eu.

Furthermore, work presented here is also linked to the research carried out by the authors at the research center on Law and Society CEDIS (Centro de I & D sobre Direito e Sociedade), at the Faculty of Law at Universidade Nova de Lisboa, in Portugal, as well as research conducted by the author at Instituto Jurídico Portucalense, at Universidade Portucalense in Porto, Portugal.

References

1. ABDI (Agência Brasileira de Desenvolvimento industrial). Nanotechnology primer (2010). ABDI, Brasília (in Portuguese). Available at: http://www.desenvolvimento.gov.br/arquivos/dwnl_1296148052.pdf. Accessed on 02 Jun. 2020.

2. Bacchini, F. (2013) Is nanotechnology giving rise to new ethical problems? *Nanoethics*, **7**, pp. 107–119. Available at: https://doi-org.ez101.periodicos.capes.gov.br/10.1007/s11569-013-0179-1. Accessed on 09 Jun. 2020.

3. Bardin, L. (2011) *Content Analysis* (Edições 70, São Paulo) (in Portuguese).

4. de Paulo Barretto, V. (2013) *The Fetish of Human Rights and Other Issues,* 2nd Ed. (Livraria do Advogado Editora, Porto Alegre) (in Portuguese).

5. Beck, U. (2003) *A World at Risk.* Translation by Laura Castoldi (Giulio Einaudi editore s.p.s., Torino) (in Italian).

6. Beck, U. (2018) *The Metamorphosis of the World: New Concepts for a New Reality*. Translation by Maria Luiza X. de A. Borges (Zahar, São Paulo) (in Portuguese).

7. Coenen, C. (2016) Broadening discourse on Responsible Research and Innovation (RRI). *Nanoethics*, **10**, 1, Available at: https://link.springer.com/article/10.1007/s11569-016-0255-4. Accessed on: 09 Jun. 2020.

8. Commission of the European Communities. (2008) Commission recommendation: 07/02/2008 on a code of conduct for responsible nanosciences and nanotechnologies research. Brussels, 7 Feb. Available at: http://ec.europa.eu/research/participants/data/ref/fp7/89918/nanocode-recommendation_ en.pdf. Accessed on: 08 Jun. 2020.

9. Dekkers, S. et al. (2016). Towards a nanospecific approach for risk assessment, *Regulatory Toxicology and Pharmacology*, **80**, pp. 46–59.

10. Delmas-Marty, M. (2013) *Resist, Empower, Anticipate or How to Humanize Globalization* (Éditions du Seuil, Paris) (in French).

11. Engelmann, W. (2007) *Natural Law, Ethics and Hermeneutics* (Livraria do Advogado Editora, Porto Alegre) (in Portuguese).

12. Engelmann, W. (2009). Human rights and nanotechnologies: In search of regulatory frameworks, *Magazine IHU*, **7**, 123, (in Portuguese).

13. Engelmann, W. (2015). Nanotechnologies as a factor of democratic approximation of the countries of Latin America: In search of regulatory molds. In: Engelmann, W. and Spricigo, C. M. (Eds.) *Democratic Constitutionalism in Latin America: Challenges of the 21st Century* (Multideia, Curitiba) (in Portuguese).

14. Falinski, M. M. et al. (2018). A framework for sustainable nanomaterial selection and design based on performance, hazard, and economic considerations, *Nature Nanotechnology*, **13**, pp. 708–714.

15. Finnis, J. (2011) *Natural Law & Natural Rights*, 2nd Ed. (Oxford University Press, Oxford).

16. von Hohendorff, R. (2015) Nanotechnological revolution, risks and reflexes in Law: The necessary contributions of transdisciplinarity. In: Engelmann, W. and Wittmann, C. (Eds.) *Human Rights and New Technologies* (Paco Editorial, Jundiaí/SP) (in Portuguese).

17. Junges, J. R. (1999–2002) *Bioethics: Perspectives and Challenges* (Unisinos, São Leopoldo) (in Portuguese).

18. Luhmann, N. (2006) *Sociology of Risk*. Translation by Silvia Pappe, Brunhilde Erker, Luis Felipe Segura. (Universidad Iberoamericana, Mexico) (in Spanish).

19. Nanotechnology Products Database (NPD). Available at: https://product.statnano.com. Accessed on 05 Jun. 2020.

20. NIOSH (2018). Revised External Review Draft - *Current Intelligence Bulletin: Health Effects of Occupational Exposure to Silver Nanomaterials*. By Kuempel, E. D., Roberts, J. R., Roth, G., Zumwalde, R. D., Drew, N., Hubbs, A., Dunn, K. L., Trout, D., Holdsworth, G. Cincinnati,

OH: U.S. Department of Health and Human Services, Centers for Disease Control and Prevention, National Institute for Occupational Safety and Health. Available at: https://www.cdc.gov/niosh/docket/ review/docket260a/pdfs/260-A-Draft-Silver-NM-CIB_8-24-18_1.pdf. Accessed on 08 Jun. 2020.

21. The National Nanotechnology Initiative - Supplement to the President's 2020 Budget (2019). Available at: www.nano.gov. Accessed on 08 Jun. 2020.

22. Toumey, C. (2019) Early voices for ethics in nanotechnology. *Nature Nanotechnology*, **14**, pp. 304–305. Available at: https://doi-org.ez101. periodicos.capes.gov.br/10.1038/s41565-019-0422-1. Accessed on 09 Jun. 2020.

23. Toumey, C. (2019) Later voices on ethics in nanotechnology. *Nature Nanotechnology*. **14**, pp. 636–637. Available at: https://doi-org.ez101. periodicos.capes.gov.br/10.1038/s41565-019-0502-2. Accessed on 09 Jun. 2020.

24. Popa, F., Guillermin, M., and Dedeurwaerdere, T. (2015) A pragmatist approach to transdisciplinarity in sustainability research: From complex systems theory to reflexive science, *Futures*, **65**, pp. 45–56.

25. Sunstein, C. R. (2019) Ruining popcorn? The welfare effects of information, *Journal of Risk and Uncertainty*, **58**, pp. 121–142.

26. Schwab, K. (2016) *The Fourth Industrial Revolution*. Translation by Daniel Moreira Miranda (EDIPRO, São Paulo) (in Portuguese).

27. Schwab, K. and Davis, N. (2018) *Applying the Fourth Industrial Revolution*. Translation by Daniel Moreira Miranda (Edipro, São Paulo) (in Portuguese).

28. Separata C. (2019) *Benefits and Risks of Nanotechnologies* (Cnpem, São Paulo) (in Portuguese). Available at: https://lnnano.cnpem. br/wp-content/uploads/2019/10/SEPARATA-CNPEM-02_ Benef%C3%ADcios-e-riscos-das-nanotecnologias.pdf. Accessed on 05 Jun. 2020.

29. Swierstra, T. and Rip, A. (2007) Nano-ethics as NEST-ethics: Patterns of moral argumentation about New and Emerging Science and Technology, *Nanoethics*, **1**, pp. 3–20.

Chapter 8

Economics of Nanotechnology-I and Modern Policy and Decision-Making about Nano

Gustavo Marques da Costa[a] and Chaudhery Mustansar Hussain[b]

[a]Instituto Federal de Educação Ciência e Tecnologia Farroupilha (IFFar)
Campus Santo Augusto, Santo Augusto, CEP 98590-000, RS, Brazil
[b]Department of Chemistry and Environmental Science,
New Jersey Institute of Technology, Newark, NJ, United States
markesdakosta@hotmail.com

8.1 Introduction

In the quest for advancing technology, there is a need for smaller, light, portable, economically viable, and reliable devices. And so, the interest in nanotechnology was born. Currently, the main objective is the synthesis and deep understanding of the physical and chemical properties of semiconductor nanostructures and nanopharmaceuticals. In regions with more advanced technology, such as the United States and Europe, there is a greater concern

Environmental, Ethical, and Economical Issues of Nanotechnology
Edited by Chaudhery Mustansar Hussain and Gustavo Marques da Costa
Copyright © 2022 Jenny Stanford Publishing Pte. Ltd.
ISBN 978-981-4877-76-3 (Hardcover), 978-1-003-26185-8 (eBook)
www.jennystanford.com

for the development of nanotechnology. In underdeveloped and developing countries, the high cost and risk associated with nanotechnological research make the private sector unable to guarantee the return on investment in the short and medium terms. Therefore, there is a low investment on the part of the industry in the necessary basic research, leaving the costs to the government. But in the United States, nanotechnology has become one of America's investment priorities, with several agencies investing in this area. Thus, the private sector invests at a level equal to or higher than that of the federal government. The basic concern in the production of these materials is to have information about their properties from the beginning, to try to prevent adverse effects, before commercialization. The creation of a database has been helping this nanotechnological development. It serves as a screening tool for nanomaterials, evaluating properties such as size, shape, toxicity, and chemical and biological activities. It is known that some synthetic nanomaterials, in contact with living tissues, cause inflammation, damage tissue and DNA, and later even result in tumors. So having public data before development facilitates the process. Even more, as the area does not have specific international regulations, regulatory agencies make them according to the type of product. International regulatory agencies have been successful in implementing procedures safely. However, the long-term use of nanoproducts for human health and the environment is not known. Ethical, economic, legal, social, toxicological, and environmental issues have hindered the applications of nanotechnology. What is being sought at this moment is the establishment of a production chain based on nano-security guidelines. For the possibility of saving energy and protecting the environment, using less scarce raw materials is a concrete step in the development of nanotechnology.

8.2 Public and Private Investments in Nanotechnology

Nanotechnology works with objects on a nanometer scale in a minuscule dimension, about 1 billion times smaller than the meter. One of the basic principles of nanotechnology is the construction of structures and new materials from atoms. In this sense, atoms behave differently and may have new properties: Making them more resistant or malleable, being able to conduct heat and electricity, being more reactive, changing color, and other diverse phenomena. In this sense, the global nanotechnology market projected in 2019 an increase of $64.2 billion, with annual growth rates of 19.8%, impacting all sectors of the economy, such as biomedicine, electronics, energy, and environment [1].

Developed countries like the United States have been creating a partnership between public and private institutions for the creation and placing on the market of nanotechnologies. In developing countries such as Brazil, much of the research is found within universities with government investments or by private, but international companies headquartered in the country. Funding usually comes from institutions made up of large business owners who aim to profit after the development of nanotechnology and research grants funded by governments. And a difference that is seen between the richest and the least favored countries is that in large nations, the search is effective in production, in the profitable benefits that these can bring and the poorest would be in the consequences that the nanos can bring to health, the environment, and well-being. The countries that currently invest the most in nanotechnology are the United States, England, Germany, and Japan. And countries like Russia, China, India, and Brazil have been making significant investments in the sector in recent years. In Brazil, it is estimated that investments exceed $2 billion per year [2, 3]. Thus, the use of patents in nanotechnology is a valid instrument for the analysis of potential risks to public and environmental health in Brazil. Therefore, patents involve drugs, aesthetic products, and food substances. However, patents are usually not resident in the country of origin and belong to legal entities.

The research in nanotechnology is mainly focused on three areas, with health being highlighted for the creation of mainly anti-tumor drugs; we still have sustainable energy, less use of coal and oil, and use for cleaning water. We can also highlight the issue of agriculture and food production, in which we are already consuming without receiving information, as well as in aesthetics, as an example we have sunscreens and the industry also with carbon nanotubes, seeking to produce the smallest equipment possible and with greater yield and storage. We cannot fail to mention information and communication technologies. Also, there is research involving the use of silver nanoparticles, which contain bactericidal activity, for the preparation of a material capable of coating wood and wood panels to be used in the manufacture of hospital furniture [4].

In this scenario, the infrastructure for research is diversified among nations. In developing or underdeveloped countries, production is more retained within universities, and in developed countries, within large companies, groups formed by multinationals in addition to universities.

Most nanotechnology publications are focused on health as mentioned earlier, mainly related to its toxicity. Research in the field is not only interesting but is also in advanced stages. The following are highlighted current nanotechnological studies with various employments, ranging from cancer treatment to energy production: a battery that recharges in 2 min, nanotechnological membrane to filter water, and gelatin nanoparticles in the brain [5].

Researchers at the Nanyang Technological University (NTU), Singapore, created a battery that recovers 70% of its charge in just 2 min. The advance is in the replacement of graphite, traditionally used in the battery anode, with a gel made with titanium dioxide nanotubes. Another creation by the Nanyang Technological University, made through the startup Nano Sun, is a water filtration membrane based also on its patented titanium dioxide nanotechnology. The material's nanoparticles destroy bacteria and break down organic compounds with the help of sunlight or ultraviolet rays, thereby purifying wastewater. Another research in the health area involved a medical student and a member of the bioengineering department at the University of Illinois, the United States, who developed gelatin nanoparticles capable of safely transporting molecules of drugs introduced via the nose to the brain, and this new method is indicated for stroke patients, who would not need to take injections.

8.3 Nanotechnology and Unintended Consequences

The total long-term consequences of the use of nanotechnologies are not yet predictable. There are studies of some substances such as nanometals, such as zinc oxide, demonstrating toxicity, but we still do not have enough research to have a position in relation to this compound in terms of damage to both living organisms and humans and the environment. In this sense, the invisible and harmful dangers to health and the environment are still little known. Currently, it is known by science that there are two categories of nanoparticles, one called biodegradable and the other non-biodegradable. The first corresponds to the group of substances classified as safe, while the second represents a risk of contamination. In this sense, what occurs is bioaccumulation in human beings, as they ingest the particles or even absorb by breathing or through skin, but do not eliminate them in their entirety in the body. In this way, these particles accumulate in the organs, and these elements unknown to the body cannot be destroyed by the immune system [6]. However, what is known is that the size of nanoparticles facilitates their propagation and transport in different ecosystems. Consequently, they tend to hinder their removal because they require more expensive and specific techniques. Also, despite economic interests and government strategic plans to encourage scientific research in the areas of synthesis, characterization, and applications of nanoparticles, there is no substantiated relationship between the results of in vitro toxicological tests and the medium-term effects of exposure to nanoparticles.

Therefore, there are still risks associated with nanotechnology, such as those related to intellectual property rights, political risks to the impact on the economic development of countries and regions, risks to privacy, risks to the environment, and risks regarding the safety of workers and consumers in contact with nanomaterials.

8.4 Nano-Mechanisms as Unique Governance Challenges: Nanotechnology and Security

Governments are finding it difficult to create laws for the development of nanomaterials, due to the great diversity of products that can arise from this technology. What is happening is a developer relying on the development of another that has already been produced and approved by some regulatory institute. However, researchers want and expect global standards and laws, because creating and launching in this area move and grow as if it were in the dark, without knowing whether it can work and even be used.

Representatives of various organizations, governmental and nongovernmental, and scientific groups, at the national and international levels, express doubts regarding the regulation of research standards, methods, tests, protocols, and how to assess the toxicity caused by nanotechnology. For the transition between the technologies that we use and the new nanotechnologies, there is a long and proven process in which there is no international regulation, but rather regulatory institutes from each country, which evaluate these products with the same rules as similar products that are already in the market. The advances in this area can solve three important problems with wide social impacts: the energy crisis, the need for better medical treatments, and the demand for clean water. However, using nanotechnology to improve living conditions can only become a reality if researchers define a safe path. The safety in the use of nanotechnology is being discussed a lot, both for the use of humans and its release in the environment [7].

In this sense, the need arises for the society to debate about the uncertainties of nanotechnology and its possible consequences on human health, with a special dedication to workers' health, as well as to identify the existing risks and ways of preventing the potential damage to the environment [8].

In this sense, immediate and especially long-term damage to health is the main focus of nano-security. Due to their size, nanomaterials have unique physical properties that can influence their absorption, distribution, and behavior in the body, in addition to being difficult to detect and control. When considering the atomic level, living matter is not distinguished from non-living, so living organisms would not be able to reject inanimate matter. Thus, the lack of experience with these materials, few exposure assessments, ignorance of toxicology, and difficulties in classifying their risks make, in this context, the use of nanotechnology somewhat unsafe, especially for human health. However, to assess exposure, it is recommended to measure beyond the basic parameters of classic toxicological doses, plus a metric dose.

There are currently hundreds of nanomaterials in consumer products on the market in several countries, without governments having promoted public dialogues. However, these omissions should be viewed with great caution. Human beings can have a long life, so the long-term impact is a priority before the release of these products for daily use.

8.5   Planning for the Unexpected

The field of nanotechnology, being recent, does not provide us with predictions about benefits and harms. Of course, we hope for the best, for profit and health and environmental benefits. Therefore, scientists, engineers, and all professional developers of nanomaterials plan for the unexpected.

For this field to be established, the criteria of size, shape, surface area, structure, construction of new detection instruments for monitoring, and adequate characterization of nanomaterials, as well as the processes that happen on the surface of the nanoparticle, when in contact with living systems, must be known. All this should have control and risk management in the entire production chain.

The pertinent issue is that innovation in nanotechnology applications is ahead of regulatory policy, leading to concerns about ethical, economic, legal, social, toxicological, and environmental issues that need to be aligned.

Therefore, the general question surrounding the success with the implementation of nanotechnology in our daily lives is certainly the consequence that this technology can bring to human and environmental health. There is also a need for the establishment of a scientific database capable of supporting the creation of a legal framework for the use of nanotechnology in various areas, consequently providing the necessary legal support.

Therefore, more studies on nanotechnology are needed, as well as knowledge about the risks that may be associated with the production and use of this technology.

8.6   Vision of Modern Life with Nano

The idea of nanotechnology was developed and popularized by Eric Drexler Kim through several articles and a book (*Creative Vehicles: The Arrival of the Nanotechnology Era*) published in 1986 [9]. However, researchers and entrepreneurs are calling for international laws on issues involving nanotechnology to put on the market what they are developing, but in reality, we see the attempt to match these products to those already existing, making them pass through regulatory companies not aware of the manufacture and consequences of these products. So these products are entering the market without everyone's knowledge. There is also how much work it will be to create laws, inspections of all these products because practically everything can be done based on nanotechnology. It will have to be a work of cooperation and detailed, requiring the publication and dissemination of the entire process to the population.

What needs to happen is the power of accountability for companies, and they also need to be held accountable individually. It is also worth mentioning that laws are supporting the use of nanotechnology in each country. In this sense, for nanotechnology to reach its full potential, a balance between risk regulation and the development of technology is necessary for the benefit of the population. However, it is also necessary to have a specific regulatory framework to establish definitions of nanomaterials and to characterize and assess their safety. Therefore, the first initiatives to obtain information on the areas that need to be further investigated to assess the harmful effects of nanomaterials on health and the environment were carried out by the Royal Society of the United Kingdom and the Royal Academy of Engineering in a study involving environmental risks to health, safety and even ethical and social implications associated with the development of nanotechnology [10, 11].

In this sense, the NANoREG project was created on an international scale, which deals with international nanotechnology regulations. It has as its members the main global regulatory bodies, such as the Organization for Economic Co-operation and Development (OECD), the International Organization for Standardization (ISO), and the European Chemicals Agency (ECHA) (Nanoreg 2018). Therefore, this project aims to provide a set of tools for risk assessment and decision-making tools in the short and medium term through data analysis and risk assessment, including exposure, monitoring, and control, to a selected number of nanomaterials already used in products; develop, in the long term, new testing strategies, adapted to a large number of nanomaterials, where many factors can affect their environmental and health impact; establish collaboration between governments and industry on the knowledge needed to properly manage risks and provide the basis for conventional approaches, mutually acceptable data sets and risk management practices [12].

In this sense, for the regulation of a nanomaterial, the USEPA requires manufacturers to provide scientific evidence that its use does not cause environmental damage and risks to public health. Thus, specific studies are needed on the behavior that each type of nanomaterial presents.

Also, the ISO certification for nanomaterials covers the standardization of three fundamental aspects: (1) terminology and nomenclature, (2) characterization, and (3) health, safety, and environmental risk assessment.

Therefore, as nanotechnology revolutionizes society, introducing new products and processes, it also brings an increasing concern with the risks associated with its different uses. Therefore, we need to learn more about how nanomaterials are changing the way we create products in the modern industry [13, 14].

In this scenario, everything that occurs in the best possible way will have more efficient drugs, as their penetration and targeting will be better. All technologies will be lighter and more powerful in terms of speed and storage. We will have recyclable, less polluting energy. Constructions will be faster and have a longer duration. The companies that invest in nanotechnology will have a high profit, being able to invest more and more, but we will also witness a huge inequality as only the richest will be able to acquire and benefit from the nanos. This will take time to reach the least favored who will become poorer and have less access to what can save their lives.

We can still highlight biomedical science, which is one of the main areas that most benefits from progress in nanotechnology. Nanoparticles have shown promise in drug transport, cancer treatment, neuroprotection, and tissue engineering. However, studies reveal that nanoparticles are being used in everything (beers, baby drinks) without knowing their safety. Also, to eliminate microorganisms, a wide variety of kitchen and cleaning items have nanoscale silver particles and none of these products have on their labels the warning of the presence of nanoparticles, or that have passed safety tests in government agencies.

Thus, humanity will learn how to use information and nanotechnology to propagate democracy and achieve a new quality of life.

8.7 Nanotechnology and Its Interfaces

The quantity of products containing nanotechnology has been growing, on average, by over 20% during the last few years according to the inventory of consumer products with nanotechnological components available on the world market [15]. The increased use of nanoparticles and nanomaterials in agriculture and industry, among other sectors, may increase, through ingestion or contact, the potential for contamination of living organisms [16].

Nanotechnology offers the prospect of major advances that will improve the quality of life and help preserve the environment. However, the same properties that make nanomaterials attractive—such as small particle size, varied shape, and high surface area—can also be responsible for toxic effects on human health and the environment.

An important point to be highlighted is the difficulty found to compare the toxicity results of nanoscale materials available in the current literature, due not only to the wide variety of methods for synthesis and preparation of nanomaterials but also to the lack of systematic studies reporting an adequate physical–chemical characterization of the sample used in the studies.

According to REACH (Registration, Evaluation, Authorization, and Restriction of Chemicals), the safety assessment of nanomaterials should follow the risk assessment methodology adopted for conventional chemicals, which is based on the following points: (1) evaluation of effects; (2) exposure assessment; and (3) risk characterization. To assess exposure, it becomes necessary to identify all potential sources of exposure. In this sense, it is important to know the manufacturing process, the activities involved, and the different exposure scenarios, as well as to identify the most likely routes of exposure. This type of information is also relevant for deciding the appropriate testing strategy (which studies routes of administration) and making recommendations on risk prevention measures.

The risk analysis involved in the production, handling, storage, incorporation, use, and disposal of nanomaterials is a complex process since there is still a gap regarding the information on the exposure limits of most newly developed nanomaterials [17, 18].

8.8 Final Considerations

Brazilian industries use very little innovation in nanotechnology. The Technological Innovation Survey carried out by the IBGE observed that only 1.8% of Brazilian innovative companies used nanotechnology. Therefore, the development of nanotechnology has the potential to expand opportunities in various sectors of the economy, such as industry, information technology, energy, security, and transportation, whether in developing solutions that reduce the impact on the environment, treating diseases, or catalyzing necessary chemical reactions in the industry to save resources. However, the protection of public health and the safety of workers require an approach committed to critical risk research and immediate action to mitigate possible exposures until their safety is proven.

The precautionary principle, integrated into several international conventions, was described as follows: "When any activity threatens human health or the environment, precautionary measures must be taken, even when the cause and effect relationships are not fully established scientific way. With nanotechnologies there is an important element of threat, which requires preventive or precautionary actions, assigning a load of responsibility to those who carry out activities with nanotechnologies that can cause damage, who consider alternatives to their new processes and activities, and who promote the public participation in the decision processes of their applications.

The analysis of environmental risks of nanomaterials depends mainly on the regulatory framework, involving the generation of protocols, which must be based on a multidisciplinary interaction to obtain a risk assessment in the most reliable way. With the increase in research in this area, which includes the environmental monitoring of nanoparticles, it will be possible to assess the risk of contamination by these materials.

However, it is necessary to implement adequate regulation and quality control protocols that guarantee reliability in the nanotechnology production process. In this sense, Brazil has installed capacity in universities and research institutes to meet and foster demands in the area and can be competitive worldwide.

Websites

1. http://nanofutures.eu
2. https://product.statnano.com/
3. http://www.nanotec.org.uk/finalReport.htm

References

1. Market, S. (2015). Nanotechnology Market to Reach $64.2 Billion in 2019. Advanced Materials & Processes, fev. 2015. Available on <https://www.asminternational.org/c/portal/pdf/download?articleId=25986127&groupId=10192 > (Access in July 2020).

2. Rossi-Bergmann, B. (2008). A nanotecnologia: da saúde para além do determinismo tecnológico. *Cienc. Cult.*, **60**(2), pp. 54–57.

3. *Patents and Innovation: Trends and Policy Challenges*, Organization for Economic Co-operation and Development. Available from:< http://www.oecd.org/dataoecd/48/12/24508541.pdf>, 2012 (Access in July 2020).

4. UFLA-O futuro e nano. Available from: <https://ufla.br/noticias/pesquisa/12155-o-futuro-e-nano>, 2018 (Access in July 2020).

5. Techtudo – Pesquisas mais curiosas em nanotecnologia. Available from: <https://www.techtudo.com.br/listas/noticia/2015/10/veja-pesquisas-mais-curiosas-em-nanotecnologia-ja-feitas-na-atualidade.html>, 2020 (Access in July 2020).

6. Unisinos- Nanotecnologia pode trazer complicações a saúde humana e ao ambiente. Available from: <http://unisinos.br/ideiasparaoamanha/nanotecnologia-pode-trazer-complicacoes-a-saude-humana-e-ao-ambiente/>, 2020 (Access in July 2020).

7. Impacts of Manufactured Nanomaterials on Human Health and the Environment; US Environmental Protection Agency. Available from: <http://es.epa.gov/ncer/rfa/current/2003_nano.html>, 2020 (Access in July 2020).

8. Ambito jurídico- Nanotecnologia e os desafios da gestão de saúde e segurança do trabalho. Available from: <https://ambitojuridico.com. br/cadernos/direito-do-trabalho/a-nanotecnologia-e-os-desafios-da-gestao-de-saude-e-seguranca-do-trabalho/>, 2020 (Access in July 2020).

9. Feynman, R. P. (1960). There's plenty of room at the bottom. *Eng. Sci.*, **23**, pp. 22–36.

10. Hankin, S. M. and Caballero, N. E. D. Regulaç,ãᵕo da nanotecnologia no Brasil e na Uniãᵕo Europeia. Available from: <file:///C:/Users/marke/ Downloads/1-dialogos_setoriais_-_nanotecnologia_portugues%20(1). pdf>, 2014 (Access in August 2018).

11. Nolasco, L. G. and Santos, N. (2017). Avanços nanotecnológicos e os desafios regulamentares. *Rev. Fac. Direito UFMG.*, **71**, pp. 375–420.

12. Nanoreg- A common European approach to the regulatory testing of manufactured nanomaterials, Plesmanweg, 1_6, JG Den Haag, Netherlands, 2019. Available from: <http://www.nanoreg.eu/.>, 2020 (Access in July 2020).

13. Hussain, C. M. (2020). *Handbook of Functionalized Nanomaterials for Industrial Applications*, 1st Ed. (Elsevier).

14. Hussain, C. M. (2020). *The ESLI Handbook of Nanotechnology: Risk Safety, ESLI and Commercialization*, 1st Ed. (Wiley).

15. Project on Emerging Nanotechnologies—Nanotechnology Consumer Product Inventory (2018), Washington, DC, 2015. Available from: <http://www.nanotechproject.org/cpi/.> (Access in July 2020).

16. Ciência Hoje – Nanomateriais e potencias riscos à saúde. Available from <http://cienciahoje.org.br/artigo/nanomateriais-e-potenciais-riscos-a-saude/> 2019 (Access in June 2020).

17. Kandlikar, M., Ramachandran, G., Maynard, A., Murdock, B., and Toscano, W. A. (2007). Health risk assessment for nanoparticles: A case for using expert judgment. *J. Nanopart. Res.*, **9**, pp. 137–156.

18. International Labour Organization Emerging risks and new patterns of prevention in a changing world of work (2010) Available from:< http://www.ilo.org/wcmsp5/groups/public/-   --ed_protect/---protrav/---safework/documents/publication/wcms_123653.pdf>, 2010 (Access in January 2020).

Chapter 9

Economics of Nanotechnology-II

Michele dos Santos Gomes da Rosa and Maurício Machado da Rosa

Cardiovascular Centre of Universidade de Lisboa–CCUL,
Faculty of Medicine, Universidade de Lisboa, Portugal
milkasg@gmail.com

Nanotechnology is a multidisciplinary scientific field that has advanced rapidly in recent years, finding applications in various areas, from energy and electronics sectors to the pharmaceutical industry. Nanoscience and nanotechnology include science, engineering, and technology conducted at the nanoscale, which involves the manipulation of mater with at least one dimension sized from 1 to 100 nm.

There are many different fields in which nanotechnology is used, including the medical field, drug delivery, biosensors, and medical imaging, as well the fabrication of nanocircuits for increasingly smaller computer processors.

Environmental, Ethical, and Economical Issues of Nanotechnology
Edited by Chaudhery Mustansar Hussain and Gustavo Marques da Costa
Copyright © 2022 Jenny Stanford Publishing Pte. Ltd.
ISBN 978-981-4877-76-3 (Hardcover), 978-1-003-26185-8 (eBook)
www.jennystanford.com

The principle of this science is that materials on the nanometric scale may present chemical, physical–chemical, and behavioral properties different from those presented on larger scales. These properties of the nanomaterials are already being industrially exploited with the manufacture of new cosmetics, medicines, coatings, and tissues. Its ranges can assort from simple development to more complex and accurate systems for loading and releasing drugs.

9.1 Nanotechnology (Benefits and Risks)

Nanoscale materials have been used for decades in applications ranging from window glass and sunglasses to car bumpers and paints. Now, however, the convergence of scientific disciplines (chemistry, biology, electronics, physics, engineering, etc.) is leading to a multiplication of applications in materials manufacturing, computer chips, medical diagnosis and healthcare, energy, biotechnology, space exploration, security, and so on.

It is this convergence of science on the one hand and growing diversity of applications on the other that is driving the potential of nanotechnologies. Indeed, their biggest impacts may arise from unexpected combinations of previously separate aspects.

Nanotechnology offers the potential for new and faster kinds of computers, more efficient power sources, and life-saving medical treatments. Potential disadvantages include economic disruption and possible threats to security, privacy, health, and the environment.

Concerns are growing that it may have toxic effects, particularly damaging to the lungs. Although nanoparticles have been linked to lung damage, it has not been clear how they cause it.

9.2 Community Ownership

Nanotechnology raises the possibility of microscopic recording devices, which would be virtually undetectable. More seriously, it is possible that nanotechnology could be weaponized. Atomic weapons would be easier to create, and novel weapons might also be developed. One possibility is the so-called "smart bullet," a computerized bullet that could be controlled and aimed very accurately. These developments may prove a boon for the military; but if they fall into the wrong hands, the consequences would be dire.

9.3   Nano-Infrastructure

According to the United States National Nanotechnology Initiative, nanotechnology is "science, engineering, and technology conducted at the nanoscale, which is about 1 to 100 nanometers." One nanometer is a billionth of a meter, or 10^{-9} of a meter. For comparison, a sheet of newspaper is about 100,000 nm thick. Scientists are discovering that atoms and molecules behave differently at the nanoscale.

It is also a rapidly expanding field. Scientists and engineers are having great success making materials at the nanoscale to take advantage of enhanced properties such as higher strength, lighter weight, increased electrical conductivity, and chemical reactivity compared to their larger-scale equivalents.

9.4   Commercialization of Nanotechnologies

Nanoeconomics is an ally of nanoscience and the economy to accelerate the pace of technological change. In an objective way, we intend with this chapter to demonstrate the need to understand the role of nanotechnology research in economic growth or what is the mechanism by which collaboration drives innovation and competitiveness.

We will demonstrate nanotechnology-enabled applications for potential benefits such as reduced costs, less toxicity, increased operational efficiency, voltage, reduced complexity, and reliability, in this chapter.

Nanotechnology holds enormous potential for healthcare, from delivering drugs more effectively, diagnosing diseases more rapidly and sensitively, and delivering vaccines via aerosols and patches. Rich countries are investing heavily in nanotechnology for health (Hussain 2020c).

The benefits associated with nanotechnology include among others increase in yield and quality of produce in agriculture, improved cosmetic products, directed delivery of medicines, and sensor applications (Hussain 2018; Hussain 2020a; Hussain 2020b).

Pharmaceutical nanotechnology is the area of pharmaceutical sciences involved in the development, characterization, and application of therapeutic systems on a nanometric or micrometric scale.

Studies of such systems have been actively conducted in the world with the purpose of directing and controlling the release of drugs. When nanotechnology is applied to the pharmaceutical segment, it represents radical new properties that have the potential to revolutionize current drug-delivery technologies and offer many opportunities to create new delivery systems for low-soluble drugs with other technical difficulties.

The benefits are both for users and for the industry, especially the insertion of innovative products with higher therapeutic efficacy, with lower production of adverse effects and, therefore, better quality of life for patients. But mainly the feasibility of administering drugs poorly soluble in water through their incorporation into nanosystems and increased stability of certain drugs, such as peptides, proteins, and nucleic acids in biological media.

9.5 National Security/Economic Competitiveness

This technology emerged in 1969, with the initial development of microencapsulation, a technique for transforming liquids (polymers and other substances) into powders, with the size of micrometric particles.

Microencapsulation served as a model for more sophisticated techniques, now on a nanoscale, allowing the development of nano-particles. Currently, nanosystems are developed, such as liposomes and nanoparticles, and microsystems such as microparticles, multiple emulsions, and microemulsions.

Despite the functionalization of nanomaterials with specific molecules, the biodiversity and availability of nanomaterials in biological systems are still reduced. But doses are also important factors to define the point where the doses administered do not demonstrate toxic effects, or how long this nanomaterial becomes available within the body until it is completely eliminated, and how it will be eliminated. Therefore, the importance of studies of standardization and validation of nanoparticles dispersed in biological media should be encouraged and optimized.

Finally, a number of experimental methodologies and details should be taken into account in nanomaterials toxicity studies for further standardization and regulation, especially for the subjects that will be administered as a diagnosis in medicine.

Pharmaceutical nanotechnology is the area of pharmaceutical sciences involved in the development, characterization, and application of therapeutic systems on a nanometric or micrometric scale.

Studies of such systems have been actively conducted in the world with the purpose of directing and controlling the release of drugs. When nanotechnology is applied to the pharmaceutical segment, it represents radical new properties that have the potential to revolutionize current drug-delivery technologies and offer many opportunities to create new delivery systems for low-soluble drugs with other technical difficulties.

Nanoscale drug delivery systems can also be used to promote the delivery of the drug to the specific target, thus reducing unwanted toxicity. Thus, contributing to better patient adherence in clinical treatment.

9.6 Nanomedicine and Human Body

The application of nanotechnology for the treatment, diagnosis, monitoring and control of biological systems was recently called "nanomedicine" by the National Institute of Health in the United States.

The rapid advance of nanomedicine is closely related to some properties of nanomaterials which allow applications in diagnostics and therapies (Chan 2006). For the development of nanomaterials with high specificity, whatever their applications, characteristics such as stability, size dispersion, morphology, surface load, and toxicity must be well defined so that the desired results are achieved.

For a high sensitivity and selectivity application in diagnostic and therapeutic systems, nanomaterials should be combined with even more specific biomolecules.

One application of nanotechnology in medicine currently being developed involves employing nanoparticles to deliver drugs, heat, light, or other substances to specific types of cells, such as cancer cells. Particles are engineered so that they are attracted to diseased cells, which allow direct treatment of those cells. This technique reduces damage to healthy cells in the body and allows for earlier detection of disease. For example, nanoparticles that deliver chemotherapy drugs directly to cancer cells are under development. Drugs containing dendrimers for targeted delivery are also being investigated.

9.7 Economic Impact

This socio-economic promise of nanotechnology has contributed to very rapid growth in public R & D investments in this field. In fact, hardly any other technology field has benefited from as much public R & D investment globally in such a short time as nanotechnology, and private sector investment is also picking up.

Nanomedicine is not only important to Europe from the social and welfare aspects, but also for its economic potential (Farokhzad and Langer 2009). It includes all products that can be defined as "systems and technologies for healthcare, aimed at prevention, diagnosis or therapy." Little market data have been published specifically about nanomedicine at present. However, an analysis of the market segments for medical devices and drugs and pharmaceuticals gives an idea about the leverage of nanomedicine on the markets. These two market segments represented in 2003 an end user value of 535 billion of euros, of which the drugs segment is the most important, with a value of 390 billion of euros. Globally this market has been growing at a 7 to 9% annual rate, with variations according to country, technologies, and market segments (Hullmann 2006).

Nanotechnology can offer impressive resolutions, when applied to medical challenges like cancer, diabetes, Parkinson's or Alzheimer's disease, cardiovascular problems, inflammatory or infectious diseases (Barua, Datta, and Das 2020).

The ageing population, the high expectations for better quality of life, and the changing lifestyle of European society call for improved, more efficient, and affordable healthcare (Barua, Datta, and Das 2020).

Nanotechnology can offer impressive resolutions, when applied to medical challenges like cancer, diabetes, Parkinson's or Alzheimer's disease, cardiovascular problems, inflammatory or infectious diseases (Boisseau and Loubaton 2011).

Experts of the highest level from industry, research centers, and academia convened to prepare the present vision regarding future research priorities in nanomedicine (Barua, Datta, and Das 2020). A key conclusion was the recommendation to set up a European Technology Platform on nanomedicine designed to strengthen Europe's competitive position and improve the quality of life and healthcare of its citizens (Bawa 2011).

9.8   Sustainability: Environment

The medical devices market is expected to grow in value by about 9% annually at present. The introduction of novel nanotechnologies can be expected to give rise to a much higher rate, by providing innovative solutions and more precise care and new information for preventive medicine. The market can be further segmented into areas where nanomedicine might have the highest potential of penetration, such as in vitro diagnostic products, patient monitoring systems, imaging systems, or imaging contrast agents (Barua, Datta, and Das 2020).

In the medical devices market of 145 billion of euros in 2003, in vitro diagnostic systems represented 18 billion of euros, or 13% of the total. It can be expected that nanotechnology will have an impact on this expanding market in the coming years, as it offers the potential of faster and more accurate analyses of smaller and smaller samples (Bosetti and Vereeck 2011).

Medical imaging systems represent 14.5 billion of euros, or 8% of the total devices market. Imaging tools and imaging agents (including contrast media and radiopharmaceuticals) represent 4 billion of euros, or 3%. These segments will benefit from the application of techniques developed from an understanding of both materials and cellular activities at the nanoscale (Hullmann 2006).

Already the sale of tools dedicated to molecular clinical and preclinical imaging represents 0.8 billion of euros out of the 14.5 billion of euros in total, and the patient monitoring market represents 1.5 billion of euros (Hullmann 2006).

Nanomedicine can also potentially affect aspects of all medical devices, for example new materials for surgical implants, nanometric systems for monitoring cardiac activities, or minimally invasive surgery sensors (Barua, Datta, and Das 2020).

When reviewing the economic potential of nanomedicine, all the biotech companies must be considered as they are directly involved in the development of new molecules, and also in the development of new tools for accelerating the discovery of appropriate molecules. Today half of the new molecules discovered worldwide come from biotech companies (Boisseau and Loubaton 2011).

There are more than 4000 companies worldwide, with over 300 in the United States, actively working on developing drug-delivery platforms, including therapies targeted to the site of the disease, as well as drug-containing implants, patches, and gels (Barua, Datta, and Das 2020).

Europe has acknowledged strengths particularly in medical devices development and in drug-delivery research, and these are clearly areas where the establishment of a European Nanomedicine Platform would contribute to maintaining and improving European competitiveness (Malloy 2011).

Nanomaterials will play a disruptive role in next-generation thermal management for high-power electronics in aerospace platforms. These high-power and high-frequency devices have been experiencing a paradigm shift toward designs that favor extreme integration and compaction (Adlakha-Hutcheon et al. 2009).

The reduction in form factor amplifies the intensity of the thermal loads and imposes extreme requirements on the thermal management architecture for reliable operation. In this perspective, we introduce the opportunities and challenges enabled by rationally integrating nanomaterials along the entire thermal resistance chain, beginning at the high heat flux source up to the system-level heat rejection. Using gallium nitride radio frequency devices as a case study, we employ a combination of viewpoints comprising original research, academic literature, and industry adoption of emerging nanotechnologies being used to construct advanced thermal management architectures (Adlakha-Hutcheon et al. 2009).

We consider the benefits and challenges for nanomaterials along the entire thermal pathway from synthetic diamond and on-chip microfluidics at the heat source to vertically aligned copper nanowires and nanoporous media along the heat rejection pathway. We then propose a vision for a materials-by-design approach to the rational engineering of complex nanostructures to achieve tunable property combinations on demand. These strategies offer a snapshot of the opportunities enabled by the rational design of nanomaterials to mitigate thermal constraints and approach the limits of performance in complex aerospace electronics (Adlakha-Hutcheon et al. 2009).

Experts of the highest level from industry, research centers, and academia convened to prepare the present vision regarding future research priorities in nanomedicine. A key conclusion was the recommendation to set up a European Technology Platform on Nanomedicine designed to strengthen Europe's competitive position and improve the quality of life and healthcare of its citizens. This article has been extracted from the vision paper "European Technology Platform on NanoMedicine - Nanotechnology for Health" produced by the European Commission (Bawa 2011).

Environmental health and safety are subject to oversight from the EPA. While EPA is already involved in the oversight of nanotechnology and nanomaterials, it plays a limited role with respect to environmental issues raised by HSR (Melnik 2011). The approval of clinical trials qualifies as a federal action requiring compliance with the National Environmental Policy Act (NEPA), which requires preparation of an environmental impact statement with a finding of no significant environmental impact or the applicability of one of several categorical exclusions (Rahman et al. 2007).

These exclusions apply broadly to almost all drug, device, biologic, and combination product approvals and are not written with any consideration for the hazard and risks that may be posed by some nanomaterials. Indeed, a categorical exclusion applies so long as the risk assessment concludes that the expected environmental concentration of a drug will be less than 1 part per billion (Resnik 2007).

Given that nanomaterials may be more reactive than non-nanomaterials at smaller scales, this exclusion deserves scrutiny.

Therefore, the importance of studies of standardization and validation of nanoparticles dispersed in biological media should be encouraged and optimized (Melnik 2011).

Finally, several experimental methodologies and details should be taken into account in nanomaterials toxicity studies for further standardization and regulation, especially for the subjects that will be administered as a diagnosis in medicine.

References

Adlakha-Hutcheon, G, R Khaydarov, R Korenstein, R Varma, A Vaseashta, H Stamm, and M Abdel-Mottaleb. 2009. "Nanomaterials, Nanotechnology." In *Nanomaterials: Risks and Benefits*, 195–207. Springer.

Barua, Ranjit, Sudipto Datta, and Jonali Das. 2020. "Application of Nanotechnology in Global Issues." In *Global Issues and Innovative Solutions in Healthcare, Culture, and the Environment*, 292–300. IGI Global.

Bawa, Raj. 2011. "Regulating Nanomedicine-Can the FDA Handle It?" *Current Drug Delivery* **8** (3): 227–34.

Boisseau, Patrick, and Bertrand Loubaton. 2011. "Nanomedicine, Nanotechnology in Medicine." *Comptes Rendus Physique* **12** (7): 620–36.

Bosetti, Rita, and Lode Vereeck. 2011. "Future of Nanomedicine: Obstacles and Remedies." *Nanomedicine* **6** (4): 747–55.

Chan, Vivian SW. 2006. "Nanomedicine: An Unresolved Regulatory Issue." *Regulatory Toxicology and Pharmacology* **46** (3): 218–24.

Farokhzad, Omid C, and Robert Langer. 2009. "Impact of Nanotechnology on Drug Delivery." *ACS Nano* **3** (1): 16–20.

Hullmann, Angela. 2006. "The Economic Development of Nanotechnology-an Indicators Based Analysis." *EU Report*.

Hussain, Chaudhery Mustansar. 2018. *Handbook of Nanomaterials for Industrial Applications*. Elsevier.

———. 2020a. *Handbook of Functionalized Nanomaterials for Industrial Applications*. Elsevier.

———. 2020b. *Handbook of Manufacturing Applications of Nanomaterials*. Elsevier.

———. 2020c. *The ELSI Handbook of Nanotechnology: Risk, Safety, ELSI and Commercialization*. John Wiley & Sons.

Malloy, Timothy F. 2011. "Soft Law and Nanotechnology: A Functional Perspective." *Jurimetrics* **52**: 347.

Melnik, TATIANA. 2011. "Mobile Tech: Is It Right for Your Organization." *J Health Care Compliance* **13** (6): 49–52.

Paschoalino, Matheus P, Glauciene PS Marcone, and Wilson F Jardim. 2010. "Os Nanomateriais e a Questão Ambiental." *Química Nova* **33** (2): 421–30.

Pautler, Michelle, and Sara Brenner. 2010. "Nanomedicine: Promises and Challenges for the Future of Public Health." *International Journal of Nanomedicine* **5**: 803.

Rahman, Syed Ziaur, Rahat Ali Khan, Varun Gupta, and Misbah Uddin. 2007. "Pharmacoenvironmentology–a Component of Pharmacovigilance." *Environmental Health* **6** (1): 1–3.

Resnik, David B. 2007. "The New EPA Regulations for Protecting Human Subjects: Haste Makes Waste." *The Hastings Center Report* **37** (1): 17.

Chapter 10

Legalization of Nanotechnology

Maicon Artmann, Roberta Verdi, Vanusca Dalosto Jahno, and Haide Maria Hupffer

Postgraduate Program in Environmental Quality, Feevale University, RS 239, 2755, Rio Grande do Sul, 93525-075, Brazil

artmann.maicon@gmail.com

Nanoscale technologies have advanced in recent decades in several countries around the world, largely due to investments in nanotechnology, which have boosted the world market, including the United States, China, and the European Union. To accelerate the development of new products with nanotechnologies, there is a need for greater interaction and cooperation between academy, industry, and government institutions. To this end, it is important to pay more attention to regulatory initiatives for the production of nanoproducts and their use, and this chapter will analyze regulatory actions in the European Union, the United States, and China, with a view to reducing the impact on quality of life and in the environment.

Environmental, Ethical, and Economical Issues of Nanotechnology
Edited by Chaudhery Mustansar Hussain and Gustavo Marques da Costa
Copyright © 2022 Jenny Stanford Publishing Pte. Ltd.
ISBN 978-981-4877-76-3 (Hardcover), 978-1-003-26185-8 (eBook)
www.jennystanford.com

10.1 Introduction

The end of the 20th century and the beginning of the 21st century are stages of technological advances driven by nanotechnology, biotechnology, artificial intelligence, synthetic biology, internet of things, neuroscience, virtual reality, and recombinant DNA technology [1–3]. All of these technological advances have produced and are still going to have profound impacts on law and forms of regulation. One of these advances is related to nanotechnology, which is growing exponentially and incorporated in numerous products and applications that include the area of cosmetics, petrochemicals, defense, textiles, health, automotive, chemistry, remediation and environmental protection, aerospace, agriculture, packaging, food, information technology, communication, among others.

Although the regulatory issue is progressing slowly, in the last two decades, numerous regulatory initiatives have been issued to promote good practices and rules to govern behaviors related to materials, products, or processes based on nanotechnology, seeking their responsible development.

Regulation and self-regulation were developed by several actors– governmental and nongovernmental organizations–at different administrative levels. Thus, the performance of several bodies led to the evolution of a polycentric regulatory governance of nanotechnology; however, the current regulatory framework is fragmented.

In this chapter, regulatory initiatives for products and applications with nanotechnology prepared by the European Union, the United States, China, and private national and international organizations will be presented. It is also the intention to defend the creation of a coordinated global regulatory system, aiming at dialogue between all sectors of nanotechnology, so that there are no negative impacts on human health and the environment.

10.2 Regulatory and Governance Initiatives for Nanotechnology in the European Union

The increasing degree of regulatory fragmentation in contemporary global governance is a phenomenon that occupies many scholars of law, international relations, and sociology, who study its causes, consequences, and responses. The example of this is the debate about the regulation of nanotechnology, or the regulation of the use of nanoparticles. The regulation of an emerging technology with possibilities for developing new products and applications in all economic sectors is extremely complex. Some countries use legislation developed for chemicals or by international nongovernmental bodies waiting to develop the regulatory framework for nanotechnology when they have a better understanding of the behavior of nanomaterials in humans and the environment.

The European Union, China, and the United States collaborated by issuing guidelines and principles based on broad stakeholder participation. Among them are the development of codes of conduct ethics, the dissemination of good risk management practices, the adoption of safe practices, establishment of exposure limits in the workplace. Following as a guideline, the development of new products and processes with nanotechnology must take place considering the benefits for society, safety and risk management.

In the European Union, the first step toward the elaboration of a policy for nanotechnologies took place between 1998 and 2002, when the Fifth Framework Program was elaborated with guidelines and policies for research, innovation, and technological development. However, it is within the scope of the Sixth Framework Program (2002–06) that nanotechnology projects were prioritized in the Commission Communication "Towards a European Strategy on Nanotechnologies" with an integrative, responsible, and safe approach. It should be noted that the communication is not a regulation on nanotechnology within the EU, but a set of guidelines with the indication that the development of products and applications with nanotechnology must take place in a safe and responsible manner, with respect to ethical principles. Research on the safety of nanomaterials is necessary to ensure a high level of protection for public health, the environment, and the consumer, continuous assessment of potential risks at all stages of the product life cycle, in addition to providing the consumer with information on potential risks, adopt measures to ensure the health and safety of workers, promote actions in cooperation with society and with international initiatives. The document, for many times, reiterates that it is essential for the development of nanotechnology in the EU that "risks be dealt with head-on as an integral part of the development of these technologies, from their conception, throughout the research and development phase and until their commercial exploitation." And that research should be carried out to "provide quantitative data on toxicology and ecotoxicology (including data on dose response and exposure of humans and the environment), in order to allow risk assessments to be carried out and, where necessary, adjustment of risk assessment procedures" [4].

The European Group on Ethics in Science and New Technologies, which was created in 1991, has an advisory role, with a global perspective, but the concern with nanoscience began in 2003, after the Green Party discussed potential risks in the European Parliament [5]. It is well known that the discussion did not start out of a purely environmental concern, but because the EU invests considerably in technological research, in order to compete with the American market, increasing the economic movement of member countries. The European scenario was changed, research was expanded, and an integrated research space was created [6]. In any case, the EU deserves special mention for the creation of three committees in 2004: Scientific Committee on Consumer Safety (SCCS), which supervises nanomaterials in consumed products; Scientific Committee on Health and Environmental Risks (SCER), which analyzes nanotechnology in food, as well as medical, health, and environmental issues; and Scientific Committee on Emerging and Newly Identified Health Risks (SCENHIR), which analyzes methodologies for assessing risk of new technologies, such as nanotechnology [7].

The European research group responsible for ethics in nanotechnology, after research with public participation, underlined that the concern with safety in relation to the manipulation of nanotechnologies is vital. In its research results, it pointed out the need to "establish measures to verify the safety of nanomedical products" and issues of military use of nanotechnology, improving economic equity and testing on animals [8].

The action plan for nanotechnology elaborated in 2005 had a high investment in research, aiming to investigate possible negative effects on health and the environment, as well as ethics in the use of nano and general issues that can contribute to development goals of the millennium. In 2007, the European Commission accepted the first report on the implementation of the action plan. The second implementation report was adopted in 2009, stating that "efforts to address corporate and security concerns must be continuous to ensure the safe and sustainable development of nanotechnology" [9]. In the communications of the Commission "is stated clearly that nanotechnology must be developed in a responsible way, within an open debate that involves the public and that enables interested people to reach their own informed and independent judgements" [10].

Another EU advisory group is ETAG, the European Technology Assessment Group, which carries out projects on the environmental, health, and safety risks of engineered nanomaterials. In 2010, ETAG was called upon by the EU to develop guidelines and policies capable of promoting industrial acceptance of new technologies from European industries. Through its reports, the group suggests two converging views, one involving mutually enabling technologies and the other applying a culture to promote improvement [11].

The Seventh Framework Program of the European Community for research, technological development, and demonstration activities for the period 2007–13 was approved by two acts: (1) Decision n. 1982/2006/EC of the European Parliament and the Council with one of the guidelines specifically aimed at the development of new products accompanied by research to manage the risks of nanoengineered nanotechnologies and manufactured nanomaterials, impacts of nanotechnology on society and potential benefits for solving problems of the company; (2) Council Decision 969/2006/EC of 18 December 2006 on the Seventh Framework Program of the European Atomic Energy Community (Euratom) for nuclear research and training activities (2007–2011). Article 2, item I of Decision no. 1982/2006/EC, when dealing with the objectives and activities that must be carried out in transnational cooperation, expressly indicates the areas of nanosciences, nanotechnologies, materials, and new production technologies, earmarking 3.475 million euros for the development of this important technological field. The activities also included research on the impact of nanotechnology on society and the interest of nanosciences and nanotechnologies in solving problems of the society [12].

The Seventh Framework Program for Research and Technological Development involved the collaboration of 47 different scientists from partner universities and industries, including China. There are also individual agreements made by countries, such as the 2002 EU–China cooperation agreement in the field of material sciences. This agreement facilitates the participation of Chinese research organizations in European research, projects with Chinese funding and vice versa. The EU and China have a joint agreement to exchange data on safety testing in order to boost consumer safety research on nanotechnology products and applications [13].

The European Committee for Standardization (CEN), the European Committee for Electrotechnical Standardization (CENELEC), and the European Telecommunications Standards Institute (ETSI) are the three European bodies responsible for the elaboration of programs to standardize safety standards, quality, risk assessment, consumer protection, environmental protection, and governance of nanomaterials in the European Union. Under Mandate 409 (M/409), the European Commission, through the Enterprise and Industry Directorate-General, asked CEN, CENELEC, and ETSI to draw up a European standardization program for nanotechnologies and nanomaterials. In the document, there is an explicit recommendation for the three bodies to initially review the existing standards, both national and international, as well as observe the European Commission's Sixth and Seventh Program, the standards of the International Organization for Standardization (ISO), and the Organization for Economic Co-operation and Development (OECD) on nanotechnology and nanomaterials, thus aiming at harmonizing legislation and uniformity in directives to meet international trade. In sequence, they are asked to identify gaps for the development of new standards, standardization documents, and "identify the availability of stakeholders in the EEA with a view to associate them when necessary in the standardisation process" [14].

The EU's first major achievement was the creation of TC-352- CEN (European Committee for Standardization) with the participation of CEN members. Its function is to carry out standardization activities that are guided by the CEN Technical Council. In turn, standards are prepared by Technical Committees (TCs) divided by areas (scope) with the responsibility to identify standards. Initially, the work of identifying standards is carried out by national members (experts) who present the national point of view, since the adoption of this practice allows the Technical Committees in sequence to be able to make balanced decisions that reflect a broad EU consensus [15].

The second major achievement of the EU is the Code of Conduct for Nanosciences and Nanotechnology approved by the Commission on 7 February 2008. The Code of Conduct lists principles for promoting the development of nanotechnology and nanomaterials in a responsible, ethical, sustainable, and safe manner. It should be adopted by all Member States with the benefit of society as a whole as its norm. The general principles and guidelines proposed in the Code of Conduct are the result of a wide public consultation [16]. The code seeks to intervene at an earlier stage in the nanotechnology development cycle and provides for the incorporation of principles of responsibility in the research phase [17].

The seven general principles that guide the Code of Conduct for research activities in nanoscience and nanotechnology (N&N) are:

1. **Meaning:** They must be understandable to the public, respect fundamental rights, and be developed for the welfare of society and individuals.
2. **Sustainability:** They must be safe and ethical, contribute to sustainable development, and do not harm or create threats to the environment and to present and future generations.
3. **Precaution:** In view of the scientific uncertainty of potential risks to human health and the environment, the precautionary principle must guide N&N research activities.
4. **Inclusion:** Respect for the right of access to information, transparency in actions and openness to the participation of all interested parties in N&N research activities.
5. **Excellence:** Meeting the best scientific standards, good laboratory practices, and defending research integrity.
6. **Innovation:** Encouraging creativity, innovation, agility, and innovative N&N planning with growth potential.
7. **Responsibility:** Organizations and researchers are responsible for the impacts on human health and the environment of present and future generations [16].

The code seeks to intervene at an earlier stage in the nanotechnology development cycle and provides for the incorporation of principles of responsibility in the research phase [17]. In relation to nano-security, China has published 15 nanotechnology standards since the creation of its Nanotechnology Standards Committee in 2005 [18].

Another important document from the European Parliament and the Council is Regulation (EC) n. 1907/2006, which came into force in 2007 and deals with Regulation on Registration, Evaluation, Authorization, and Restriction of Chemicals (REACH). Even though not dealing specifically with nanotechnology, REACH is used to register some nanomaterials. However, the regulation is considered fragile for the registration of chemical substances and nanoparticles, which makes it impossible to have sufficient information on the dangers of certain substances to human health and the environment. REACH is based on the precautionary principle, for the assessment of all hazardous substances. Chemical safety assessments must be carried out for all substances registered in REACH. Suppliers or importers of hazardous chemical substances have an obligation to provide safety data sheets for information through risk management. Information about nanomaterials should be on the safety data sheets with a specific mention of the particles in each product [19].

In 2009, the main innovation was the approval of EC 1223/2009, which regulates the use of nanotechnology in cosmetics. The Cosmetic Regulation was the first European legal instrument to contain specific rules on nanomaterials in cosmetics [20]. The regulation also indicates what may or may not be placed in a cosmetic and requires the industry to provide information on the use of nanomaterials. Article 19 imposes that "all ingredients present in the form of nanomaterials must be clearly indicated in the list of ingredients, and the names of those ingredients must be followed by the term nano" [20].

In sequence, in 2012, the Scientific Committee on Consumer Safety (SCCS) edited SCCS guide no. 1484/2012 with guidelines for assessing the safety of nanocosmetics. As the nanocosmetics area is very dynamic, with new products tested and made available on the market, the regulation has also been updated in order to provide greater safety to consumers and workers exposed to nanocosmetics. Thus, new guides were edited. The last edition took place in October 2019, when the European Union updated the guidelines on safety assessment of cosmetics with nanomaterials SCCS/1602/18. Among the changes, there is an indication that the proposed methodology does not primarily use animals for testing. New subsections were introduced with indication of alternative methods and new methodologies of approaches for toxicological assessment of nanomaterials, assessment of the safety of the nanocosmetic for human health and the environment, procedures and schedules for notification and assessment of nanocosmetics, physicochemical characterization, risk assessment, toxicological assessment, and exposure assessment. Updates are carried out quite frequently in the European Union and always take into account scientific advances in the area, new products developed, new knowledge in the area of research, and safety of nanomaterials [21, 22].

Three European initiatives that positively call attention to the concern with responsible research and innovation, safety and legal, social, and ethical impacts of innovation are Responsible Research and Innovation (RRI); Ethical, Legal, and Social Aspects (ELSA); and Ethical, Legal, and Social Implications (ELSI). It is observed that the three initiatives aim to implement responsible practices in research and innovation, which shows again that the EU is always one step ahead and that these initiatives seek to "prevent possible impacts of the development of technologies, not with the intention of blocking progress, but with the purpose of defending population from risks and possible irreversible shocks" [23].

With the recommendation that an "international welfare code" is needed in the globalized world, ELSI believes that technological development must "harmonize values of diversity, social justice, transnational security and responsibility environmental" [24].

There have been a number of EU projects aimed at increasing public dialogue on socioeconomic issues and concerns about nanotechnology, aimed at facilitating a dialogue between science, the economic system, the legal system, and civil society on their potential benefits and impacts, bringing together nanobiotechnologists, to anticipate and discuss societal and ethical aspects [25].

Responsible Research and Innovation (RRI), by fostering work and engagement throughout the research and innovation process of the various social actors (researchers, citizens, policymakers, business, third-sector organizations, etc.), assists in the alignment of inclusive practices and sustainable with society's values, needs, and expectations. In the EU Research and Innovation Framework

Program (2014–18), named Horizon 2020, which is based on three pillars—scientific excellence, industrial leadership, and societal challenges—the implementation of the RRI is promoted by the objective "Science with and for Society." In the Horizon 2020 document, producing responsible research and innovation is a transversal theme and that it is present in the specific objectives, in the actions, and in the explicit indication of taking the lead in enabling and industrial technologies [26].

Another important EU initiative is the inclusion of nanomaterials in REACH, starting in 2020 [27].

It is noted that in the EU, there are numerous instruments that serve as an information base for the responsible development of nanotechnology, as well as a large database in order to inform the risks to the exposure of nanomaterials. Several government agencies and nongovernmental research institutes work for responsibility in research and innovation and that can contribute to the transnational standardization of nanotechnology.

10.3 An Overview of Regulatory Initiatives in the United States and China

Nanomaterials have become an important part of the development of the great world powers with billionaire investments from states and companies to leverage important sectors of the economy quickly, so as not to be left out of such a promising market. In addition to the numerous benefits of this new technology, products and applications with nanotechnologies can also pose a threat to human health and the environment due to the possibility of use in the most diverse areas.

Regulatory authorities of states, economic blocs, and nongovernmental and private organizations "closely observe recent developments in this area, seeking a balance between consumer safety and the interests of the industry" [28].

At the moment, most of the countries are more attentive to observe the development of the area and to understand potential risks in the chain of each nano-product, which is why the focus is on editing policies, recommendations, guidelines, requirements to be observed, and methodologies for risk investigation, with more robust data, develop regulatory frameworks for nanotechnology. In this section, the regulatory initiatives of two countries that are among the leading countries in the nanotechnological market will be briefly presented: the United States of America and China.

In the United States, the Food and Drug Administration (FDA) is responsible for the supervision and regulation of food, medicines for human use, cosmetics, vaccines, dietary supplements, electronic products that emit radiation, veterinary products, medical devices, tobacco products, and biopharmaceuticals, with the objective of protecting public health and ensuring that they are safe, effective, properly labeled, healthy, and in good sanitary conditions. The FDA created a task force for nanotechnology and in 2014 issued the "Guidance for Industry Considering Whether an FDA-Regulated Product Involves the Application of Nanotechnology" for manufacturers, suppliers, importers, and other stakeholders. The document is not legislation and is not legally mandatory, but it does offer a series of guidelines and recommendations supported by scientific bases that aim to guide companies on potential implications of nanomaterials in terms of safety, effectiveness, risk, impact on public health, and regulatory status of the product. Consultation at the design stage is encouraged, and the FDA makes it clear that its regulatory policy is based on products and science [29].

The Environmental Protection Agency (EPA) establishes requirements for registration of chemical substances manufactured or processed at the nanoscale, including imports and those that are still in the manufacturing design phase. The EPA requires information about the specific chemical identity of the nanomaterial, methods of manufacture and processing, level of exposure to the product, production volume, and possible effects on human health and the environment. Initially, the EPA started working with specific rules for nanotechnology as part of the Toxic Substances Control Act (TSCA). In 2011, two new regulations were proposed by the Office of Chemical Safety and Pollution Prevention with the expansion of the environmental risk framework to include nanoscale chemicals [30].

In 2017, the EPA issued the EPA-HQ-OPPT-2010-0572-0137 and 92 FR 3641 directives, which establish requirements for reporting and registering substances manufactured or processed at the nanoscale. In this standard, the EPA specifically requires companies or individuals that manufacture or process nanoscale chemicals, including imports, to prepare inventory information in order to provide TSCA with up-to-date information with scientific, technical, and economic evidence to assess each chemical substance manufactured at the nanoscale, thus facilitating new EPA regulations. Another norm is the one that deals with Principles of Regulation and Supervision of Emerging Technologies and the Decision Making of the USA, with special focus on nanotechnology, aiming to facilitate the assessment of "risks and risk management, the examination of the benefits and costs of other future measures and decision-making based on available scientific evidence" [31].

On January 28, 2020, the EPA published another document in the Federal Register requesting the Office of Pollution Prevention and Toxics (OPPT) to expand the request for data on chemical substances and other nanoparticles used in products or processes to compose the inventory. This document requests information on health and environmental effects, human and environmental exposure, and information on the disposal of materials with nanoparticles. The inventory report will enable the EPA to evaluate the data collected and enact new standards or readjust standards and "measures under the TSCA to reduce any risk to human health or the environment." Whenever the information collected is not considered confidential business information, access to "environmental groups, environmental justice advocates, state and local government entities and other members of the public will have access to this information for their own use" [32].

In the United States, there is also the National Nanotechnology Initiative (NNI), created in 2001 to develop research and development policies for nanotechnology with the involvement of 20 federal departments, independent agencies, and special commissions. The strategic planning must be updated every three years when objectives and priorities are contemplated or updated for each 3-year period. One of the main objectives of NNI is the responsible development of products and applications with nanotechnology. Attention to ethical, legal, and social issues must permeate its strategic options. Likewise, strategic decisions at the NNI, which guide nanotechnology research and advances, must be transparent and responsive to ensure public confidence in nanotechnology, promote innovation and commercialization based on responsible innovation [33].

The National Institute for Occupational Safety and Health (NIOSH) is responsible in the United States for the creation, recommendation, and communication of occupational safety and health standards and for the definition of safe exposure levels in the workplace, as well as the development of research supported by "innovative methods, techniques and approaches" for the definition of labor exposure standards. In 2004, the Nanotechnology Research Center (NTRC) was created to coordinate research efforts and actions seeking to identify risks of nanotechnologies in the work environment with the participation of scientists [34].

Researchers at the NIOSH observed in their research the adverse health effects of animals exposed to nanotechnology and risks to workers exposed in the workplace. Permanent investments in research enable the creation of standards and guidelines on occupational exposure limits, the creation of new control technologies, safety equipment and medical surveillance methodologies, as well as the observation of gaps for the development of new research. In the research plan for the period 2018–25, the NIOSH has prioritized developing research and generating data to understand new nanomaterials and the risks that can cause damage to workers' health; protect workers working with nanotechnology; disseminate good risk management practices; standardize safe worker exposure limits; support epidemiological studies with medical, cross-sectional, prospective, and worker exposure studies; and promote national and international adhesion with risk management promotion and guidance actions [34].

Another key player in the development of nanotechnology is China, which ranks first in the world market and is also the world leader in peer-reviewed scientific publications on nanotechnology. China, in the last decades, has dedicated efforts to standardize and characterize nanomaterials, carrying out tests, validating methodologies, developing standards, for example, for titanium dioxide nano, rules for measuring the nanoscale, safety, quality, and inspection of products, as well as quarantine determination for workers with possible contamination by nanoparticles. It is not China's intention to expand legal requirements, but as the EU has been reacting and making a great effort to regulate nanotechnologies, China does not want to lose space in the global market and standards for standardization can be seen positively [28].

In 1986, China launched the National High Technology Research and Development Program (known as Program 863), which, among other objectives, aimed at "promoting the development of key novel materials and advanced manufacturing technologies for raising industry competitiveness," including nanomaterials. It is a research agenda that is updated every five years, with specific programs that dictate the pace of development and government investments to finance research on nanomaterials and nanostructures. One of the Programs is China's National Basic Research Program, named Program 993, which has one of the projects aimed at the "standardization of procedures and assessment/test protocols which form the basic framework for the regulation of nanomaterials" [35].

The Medium and Long-term Development Plan (MLP) prepared for the period 2006–20 shows that the goals and objectives elaborated express China's ambitions to become a global leader in nanoproducts; that is, by giving priority to development of nanotechnologies, MLP establishes itself as a "mega project" in China's science system with the active participation of universities. On the other hand, there is also a concern to minimize "negative externalities associated with accelerated growth" and to face problems related to environmental degradation and energy efficiency. Safety and handling standards, as well as exposure and toxicity standards, are continuously developed and revised by the Nanotechnology Standardization Technical Committee (NSTC) and the Technical Committee 279 nanomaterial-specific sub-committee under the Standardization Administration of China (SAC). SAC/TC279 coordinates the elaboration of mandatory standards, voluntary (or recommended) standards, and technical standardization guides, "including terminology, methodology, and safety in the fields of nanoscale measurements, materials and nanoscale biomedicine." Technical standards for nanoscale in China can be national, sectoral, local, or for company. It is also the responsibility of SAC/TC279 to prepare test protocols, methodologies, product specifications, and technical safety standards for the production, packaging, and transportation of nanomaterials, quality of nanomaterials, and minimizing impacts on the environment and worker's health and safety [35].

Based on the synthesis carried out on two of the great world leaders in nanotechnology research and development, we agree with Abbott, Marchant, and Sylvester when they observe that leaders will actively continue to develop products and applications with nanotechnologies and "do not wish to put their scientists and companies at a competitive disadvantage by unilaterally imposing restrictive rules." In this sense, it is not surprising that many initial discussions about whether or not to regulate nanotechnologies are aimed at seeking international regulation or at least international harmonization. What the authors defend is an international harmonization to regulate nanotechnology, since no country wants to lose market in such a promising field [36].

A rigid internal regulatory framework can become a barrier to the development of products and applications with nanotechnologies. Likewise, a very strict regulation by international organizations could also mean an obstacle to a market as attractive as the nanotechnology market. However, at no time can risks to human health and the environment be neglected. It is urgent to move toward responsible governance based on human rights and care in favor of present and future generations.

10.4 Contribution of Intergovernmental Organizations to the Governance or Risk Management of Nanotechnologies

The existing traditional regulatory models for chemicals used for nanotechnologies are not suitable for regulating this new technology, which is why new ways of developing regulation must be established to deal with the complex international nanotechnological chain of products and applications and, thus, ensure that nanomaterials are safe for humans and the environment. International organizations and the various stakeholders in the production chain have different understandings and divergent views on what to regulate, how to regulate, how to develop products with safety, and perceptions of risk [37].

Good governance practices in nanotechnology and regulations are currently more dictated by private international organizations such as the WTO, OECD, and ISO, which shows a tendency toward the privatization of law in this field with indications of "openness to the movement of legal pluralism" [38]. Nanotechnologies, in the scenario of economic globalization and transnational and transgenerational risks, start demanding a "Transnational Law" that, in Ramos' words, "would be neither national nor international, but the result of the concatenated action of private entities, with the direct support or indirect of States" with the objective of regulating social facts that escape the borders of states. With the new technologies, this Transnational Law also began to encompass "efforts to produce non-mandatory standards at the international level, as a result of the action of international organizations, such as 'model laws', 'conduct guides' and 'principles or rules' that would inspire private agents and states themselves" [39]. Following, some regulatory initiatives by private international organizations that contribute to the standardization of basic concepts, nomenclatures, standardization, security protocols, risk governance, and responsibility of the actors involved will be presented, as they have resulted in the collective creation of mechanisms and regulations, in the absence of regulation on nanomaterials and nanoproducts.

The OECD is an intergovernmental organization based in Paris, France, made up of 36 member countries, bringing together representatives of the most advanced economies in Europe, North and South Americas, Asia, the Pacific, and the European Commission. Its main objective is to foster international cooperation and policy harmonization to address the main political, economic, and socio-environmental challenges between member countries with other countries and with international organizations. For each strategic theme, specialized committees and working groups are organized with the participation of delegates appointed by member countries. One of the working groups is the Working Party on Manufactured Nanomaterials (WPMN), which is linked to the Chemicals Committee and the Environment, Health and Safety Division. The working group "concentrates on human health and environmental safety implications of manufactured nanomaterials (limited mainly to the chemicals sector), and aims to ensure that the approach to hazard, exposure and risk assessment is of a high, science-based, and internationally harmonized standard." In 2005, the first report on the safety of nanomaterials was launched, which gave rise to a Series on the Safety of Manufactured Nanomaterials. The aim of the series is to provide member countries and the international community with up-to-date information on the best available methodologies for conducting risk assessment tests on human health and the environment, physical–chemical parameters, and behavior of nanomaterials in biotic and abiotic systems [40].

From 2005 to 2019, 91 reports were included and published in the Series on the Safety of Manufactured Nanomaterials. It is noticed that in the documents, the safety of manufactured nanomaterials, possible risks to human health and the environment are the concerns that underpin the reports. In 2019 alone, three new reports were launched: (1) n. 89: Developments in Delegations on the Safety of Manufactured Nanomaterials - Tour de Table; (2) n. 90: Physical–Chemical Decision Framework to inform Decisions for Risk Assessment of Manufactured Nanomaterials; (3) n. 91: Guiding Principles for Measurements and Reporting for Nanomaterials: Physical Chemical Parameters. The last document, n. 91 of 2019, is in line with the Physical–Chemical Decision Framework to Inform Decisions for Risk Assessment of Manufactured Nanomaterials 2019 and the responsibility for publication lies with the OECD Joint Meeting of the Chemicals Committee and the Working Party on Chemicals, Pesticides and Biotechnology. The proposal is to be a reference document for conducting research and development of products with nanotechnology and to establish international standards based on evidence [41].

The Testing Program of Manufactured Nanomaterials and the creation of the Working Party on Manufactured Nanomaterials (WPMN) by the OECD are two initiatives considered extremely positive because they are one of the first actions involving the collaboration of governments, development agencies, private industries, and research institutions for the generation of dossiers on the best methodologies available for carrying out safely tests on each nanomaterial listed as a priority. This care reflects concerns about the safety of nanomaterials already available on the market or undergoing tests for commercial use. It is also important to note that the program was divided considering two levels of participation:

1. **Sponsors:** Assume responsibility for conducting or coordinating all of the testing determined to be appropriate for each of the endpoints for a specific nanomaterial. In some cases, "joint lead" arrangements have been developed.

2. **Contributors:** Provide test data, reference, or testing materials or other relevant information to the lead and co-sponsors. However, both sponsors and collaborators worked together in the preparation phase of the "Dossier Development Plans" (DDPs) for the testing of each listed nanomaterial, culminating in a list of the various endpoints that each material was tested against as well as sample preparation and dosimetry information [40].

The OECD estimates that the world market for nanotechnological inputs, considering nanomaterials and nanodevices, will be $24.56 billion in 2025 [27]. In 2018, the OECD published Serie n. 88 (Series on the Safety of Manufactured Nanomaterial) with the results of "Investigating the different types of risk assessment of manufactured nanomaterials" and "Identifying tools available for risk management measures and uncertainties driving nano-specific data needs." The conclusion is that, despite the variety of methodologies and tools examined to assess and manage the risks of nanomaterials, there are many gaps and the level of scientific uncertainty regarding risks is still high [42].

Another example is the International Organization for Standardization (ISO), which is an independent, nongovernmental, international nonprofit organization that issues voluntary international technical standards relevant to the market, with the purpose of standardization and support for innovation and presentation of solutions to global challenges. In 2020, it has 784 technical committees and subcommittees that are organized to develop and standardize technical norms with the addition of 23,183 international standards that cover all aspects of manufacturing and the main technologies. Regarding nanotechnology, the ISO has developed a relevant work for the standardization and manipulation of nanoparticles. The standards issued by the ISO are agreed by experts, covering a variety of activities, which run through the entire product chain, with a special focus on the development of quality management standards, risk management, environmental management standards, safety standards for IT, health and safety standards in the workplace, food safety standards, and energy management standards [43].

The various ISO standards provide forms of risk assessment for nanotechnologies. Such standards are methodological tools for the development of the nano industry. The ISO has a list of members from 164 countries, including major world powers such as China, the United States, Germany, France, Japan, among others. France was involved in establishing 737 standards, followed by Germany with 736, and China with 731 standards [26]. In 2008, the ISO published the first two standards that define the basic terms often used in the nanotechnology literature. Technical specification no. 27687/2008 provides definitions and information for nanoscale objects and covers the terms nanoparticles, nanofibers, and nanoplates. The second Technical Report is n. 12885/2008 on health and safety practices in occupational environments relevant to nanotechnologies, which provides companies, researchers, workers, and others with information relevant to adverse health and safety consequences during the process of production, use, and disposal of manufactured nanomaterials. The largest group of ISO works is the terminology and nomenclature. This group is responsible for "defining and developing unambiguous and uniform terminology and nomenclature in the field of nanotechnologies to facilitate communication and promote a common understanding" [44].

Although there is still no legislation capable of regulating the manipulation of nanoparticles, due to the complexity of being an emerging technology, the ISO has shown to be attentive to the development of products and applications with nanotechnology and has been seeking, through discussions and certifications, to support the market in a responsible and sustainable way [45]. By May 2020, 113 standards and projects were edited under the direct responsibility of the ISO/TC 229 Technical Committee that regulates nanotechnology. As there are countless, some representatives for the present study will be mentioned:

1. TC 229 Technical Committee - Nanotechnologies; ISO/TR 12885: 2008 - Nanotechnologies-Health and safety practices in occupational environments relevant to nanotechnologies;
2. ISO/TR 13121: 2011, which deals with Nanotechnologies - risk assessment of nanomaterials;
3. ISO/TS. 18401: 2017, which deals with Nanotechnologies - Explanation in simple language of the ISO/IEC 80004 series;

4. ISO/TS 12901-1: 2012, which deals with Nanotechnologies - Occupational risk management applied to engineering nanomaterials - Part 1: Principles and approaches;

5. ISO/TS 12901-2: 2014, which deals with Nanotechnologies - Occupational risk management applied to engineering nanomaterials - Part 2: Use of the control strip approach;

6. ISSO/TS 80004-1: 2015, deals with Nanotechnologies - Vocabulary - Part 1: fundamental terms;

7. ISO/TS 80004-2: 2015 on Nanotechnologies - Vocabulary - Part 2: Nano-objects;

8. ISO/TR 18196: 2016 provide on Nanotechnologies - Measurement technique matrix for the characterization of nano-objects;

9. ISO/TR 18637: 2016, Nanotechnologies - Overview of the structures available for the development of occupational exposure limits and bands for nano-objects and their aggregates and clusters (NOAAs);

10. ISO/TS 19807-1: 2019: "Nanotechnologies - Magnetic nanomaterials - Part 1: Specification of characteristics and measurements for magnetic nanosuspensions";

11. ISO/TS 19808: 2020: "Nanotechnologies - Carbon nanotube suspensions - Specification of characteristics and measurement methods";

12. ISO/TS 21412: 2020: "Nanotechnologies - Nano-object-assembled layers for electrochemical bio-sensing applications - Specification of characteristics and measurement methods";

13. ISO/TS 22082: 2020: "Nanotechnologies - Assessment of nanomaterial toxicity using dechorionated zebrafish embryo" [46].

For developing countries, the adoption of ISO standards represents an important source of technological know-how, standardization, and access to global markets, "as they define characteristics that products and services must fulfill for export markets, representing a fair access path for fair participation in international trade" [47]. Adopting ISO standardization gives security to the entrepreneur, as they are constantly expanded and updated, which is important for the economically growing nanotechnology market. All extensions are based on environmental principles, so that the market respects sustainability and verifies the risk of the measures [48]. Thus, the "ISO seeks to provide a method of standardization in specifications and procedures, for worldwide use, through the standards it publishes" [45].

Snir and Ravid, when analyzing various nano-specific regulatory initiatives, concluded that when international standards were established, organizations became "information centers," which play a strategic role as intermediaries, spreading out as national and global policies. Through this process, these centers help to shape supranational policies. These understandings about the role of international organizations end up setting private standards and, therefore, shed new light on the debate. However, the authors note that less attention has been paid to studies on how organizations and regulatory arrangements interact in the absence of supervision by a transnational coordinating organization. "It is not fragmentation itself, but the coordination (or lack of) of fragmented or differentiated institutions that make the issue more important." Therefore, understanding the nature of continuous governance interaction is a key to any assessment of the effects of fragmentation [49].

Regulatory initiatives by governmental and nongovernmental organizations to establish behaviors related to nanomaterials have proliferated in the past decade, contributing to what Snir and Ravid call the phenomenon of pluralist regulatory governance. However, much of the regulation was "developed without guidance from international organizations, such as the United Nations Environment Program (UNEP) or the World Health Organization (WHO), which only recently became involved in the field," which resulted a "dense and polycentric regulatory governance." This is a practice observed for several products and represents a "transformation of governance, which transfers the authority to regulate environmental, health and safety problems from formal State institutions to self-organized and interorganizational networks" [49].

In all the analyzed initiatives, the concern of international organizations with the responsible development of nanotechnologies and the indication of the adoption of safe practices to minimize the risks that nano can cause to human health and the environment is perceived. So standardization or governance in nanotechnology has points in common between the international organizations studied, the European Union, the United States, and China, which demonstrates that the leaders in research and development of nanomaterials are also those who promote international standardization, occupying a privileged place in the nanotechnology scenario.

The topology of the structure of international organizations and the dialogue between public and private actors facilitate the development of regulations. Regulating nanotechnologies becomes more and more complex and global, and to move forward, it is necessary to reflect on possibilities for global governance. Finally, the network centrality measures can identify the elite group of actors in the network based on their position, and focus attention on a small number of key organizations and regulatory initiatives.

10.5 Conclusion

This study aimed to show that an anticipatory global governance, with the direct participation of public and private actors, enables new forms of society self-regulation in the face of the pressing need for responsible governance with directives for the production and safe use of nanotechnologies.

Building a new architecture for the regulation of complex topics such as nanotechnologies is a challenge, compounded by the diversity of products and applications produced and made available on the global market, by the characteristics of pervasiveness (entering into various sectors), solubility, variations in shape and size, bioaccumulation, crystalline structure, agglomeration, and scientific uncertainty regarding risks.

The fluidity of international trade in relation to nanotechnology products and applications is a reality. It is clear that each state must adopt an internal regulatory framework for responsible management of the risks of nanotechnologies to human beings and the environment. However, it is understood that an anticipatory, responsible, transdisciplinary, and global governance can be carried out if it is based on transnational and trans-temporal risk management, care ethics, nano-security protocols, responsible innovation, dialogue, and the active participation of transnational organizations, the UN, and the legal, economic, social, and political system.

Websites

1. https://eur-lex.europa.eu
2. https://epa.gov
3. http://www.oecd.org
4. https://www.fda.gov
5. http://www.safenano.org
6. https://www.iso.org/obp/ui/#iso:std:iso:ts:27687:ed-1:v2:en

References

1. Abbott, K. W, Marchant, G. E., and Sylvester, D. J. (2010). Transnational regulation of nano-technology: Reality or romanticism? In: Hodge, G. A., Bowman, D. M., and Maynard, A. D. (Eds.), *International Handbook on Regulating Nanotechnology*. Cheltenham (Edward Elgar), pp. 525–526.

2. Hussain, C. M. (2018). *Handbook of Nanomaterials for Industrial Applications*, Elsevier.

3. Hussain, C. M. (2020). *Handbook of Manufacturing Applications of Nanomaterials*, Elsevier.

4. Hussain, C. M. (2020). *Handbook of Functionalized Nanomaterials for Industrial Applications*, Elsevier.

5. European Commission (2014). Programa-quadro de investigação e inovação da UE. HORIZON 2020 em breves palavras. Luxemburgo: Serviço das Publicações Oficiais das Comunidades Europeias, pp. 5–8.

6. Mnyusiwalla, A., Daar, A. S., and Singer, P. A. (2003). Mind the gap: Science and ethics in nanotechnology. *Nanotechnology*, pp. 09–13.

7. Pandza, K., Wilkins, T. A., and Alfoldi, E. A. (2011). Collaborative diversity in a nanotechnology innovation system: Evidence from the EU Framework Programme. *Technovation*, **9**, pp. 476–489.

8. Decker, M., Li, Z. (2009), apud Dalton-Brown, S. (2015). *Nanotechnology and Ethical Governance in the European Union and China Towards a Global Approach for Science and Technology*. Springer, Melbourne, pp. 90–95.

9. University of Rochester Medical Center Rochester (2006). Tiny inhaled particles take easy route from nose to brain. Newsroom. August 03, 2006. At: <https://www.urmc.rochester.edu/news/story/1191/tiny-inhaled-particles-take-easy-route-from-nose-to-brain.aspx>.

10. Doubleday, R. (2011). Risk public engagement and reflexivity: Alternative framings of the public dimensions of nanotechnology. *Health, Risk & Society*, **9**, pp. 211–227.

11. Hullmann, A. (2008). European activities in the field of ethical, legal and social aspects (ELSA) and governance of nanotechnology. *Nano and Converging Sciences and Technologies*, pp. 3.

12. Cameron (2007) apud Dalton-Brown, S. (2012). Global ethics and nanotechnology: A comparison of the nanoethics environments of the EU and China. *NanoEthics*, **2**, pp. 137–150.

13. European Union (2006). Decision no 1982/2006/EC of the European Parliament and of the Council. *Official Journal of the European Union.*

14. Jarvis, D. and Richmond, N. (2010), apud Dalton-Brown, S. (2012). Global ethics and nanotechnology: A comparison of the nanoethics environments of the EU and China. *NanoEthics*, **2**, pp. 137–150.

15. European Commission (2007). Enterprise and Industry Directorate-General. Competitiveness in the Pharmaceuticals Industry and Biotechnology Brussels. Mandate/409. At: <https://ec.europa.eu/growth/tools-databases/mandates/index.cfm?fuseaction=search.detail&id=369#>.

16. European Committee for Standarization (2020). CEN/TC 352-Nanotechnologies. At: <https://standards.cen.eu/dyn/www/f?p=204:7:0:::: FSP_ORG_ID:508478&cs=1A6FDA13EC1F6859FD3F6 3B18B98492ED>.

17. European Commission (2009). Directorate-General for Research. Science, Economy and Society. Commission recommendation on a code of conduct for responsible nanosciences and nanotechnologies research & Council conclusions on Responsible nanosciences and nanotechnologies research. Luxembourg: Office for Official Publications of the European Communities. At: <https://ec.europa.eu/research/science-society/document_library/pdf_06/nanocode-apr09_en.pdf>.

18. Hu, W. and Chen, K. (2004). Can Chinese consumers be persuaded? The case of genetically modified vegetable oil. *AgBioForum*, **7**, pp. 124–132.

19. European Commission (2019). Environment. At: <http://ec.europa.eu/environment/chemicals/reach/reach_en.htm>.

20. European Union (2006). Decision no 1982/2006/EC of the European Parliament and of the Council. *Official Journal of the European Union.*

21. Castillo, A. M. P. (2010). The EU approach to regulating nanotechnology. Brussels: European Trade Union Institute. At: <file:///F:/

ORIENTA%C3%87%C3%95ES%20DOUTORADO%20E%20
MESTRASO/Orienta%C3%A7%C3%A3o%20doutorado/ROBERTA/
Nano-working-paper.pdf >.

22. European Union (2019). European Union Policy Document. Scientific Committee on Consumer Safety. Guidance on the Safety Assessment of Nanomaterials in Cosmetics. At: <https://ec.europa.eu/health/ sites/health   /files/scientific_committees/consumer_safety/docs/ sccs_o_233.pdf>.

23. Hussain, C. M. (2020). *The ESLI Handbook of Nanotechnology: Risk Safety, ESLI and Commercialization*, 1st Ed., Wiley.

24. Hohendorff, R. (2018). A Contribuição do Safe By Design na Estruturação Autorregulatória da gestão dos riscos nanotecnológicos: Lidando com a Improbabilidade da Comunicação Inter-Sistêmica entre o Direito e a Ciência em Busca de Mecanismos para Concretar os Objetivos de Sustentabilidade do Milênio, pp. 63.

25. Chunliang, F. (2010) Ethical environment of nano-science— Presentation Transcript. In: *The 3rd International Workshop on Innovation and Performance Management*, 1–4. University of Kent, pp. 1–29. At: <https://pt.slideshare.net/KentBusinessSchool/paper-4- ethical-environment-of-nanoscience-chunliang-4684190>.

26. Friedman, B. (2010), apud Dalton-Brown, S. (2012). Global ethics and nanotechnology: A comparison of the nanoethics environments of the EU and China. *NanoEthics*, **2**, pp. 137–150.

27. International Organization for Standardization (2020). ISO Members. At: <https://www.iso.org/members.html>.

28. Brazil (2019). Ministério da Ciência, Tecnologia, Inovação e Comunicação. Centro Nacional de Pesquisa em Energias e Materiais (CNPEM). Benefícios e Riscos das Nanotecnologias. (CNPEM, Brasília).

29. Wacker, M. G. and Proykova, A. (2016). Dealing with nanosafety around the globe: Regulation vs. innovation. *International Journal of Pharmaceutics*, **1**.

30. United States of America (2014). Department of Health and Human Services. Food and Drug Administration (FDA). Office of the Commissioner. Guidance for Industry Considering Whether an FDA-Regulated Product Involves the Application of Nanotechnology. At: < https://www.fda.gov/media/88423/download >.

31. Environmental Protection Agency (2014). Nanoscale Materials; Chemical Substances When Manufactured, Imported, or Processed as Nanoscale Materials; Reporting and Recordkeeping Requirements;

Significant New Use Rule (2070-AJ54). At: <http://www.epa.gov/oppt/nano/>.

32. Environmental Protection Agency (2017). Chemical Substances When Manufactured or Processed as Nanoscale Materials; TSCA Reporting and Recordkeeping Requirements. ID: EPA-HQ-OPPT-2010-0572-0137. *Federal Register*, **8**. At: < https://www.regulations.gov/document?D=EPA-HQ-OPPT-2010-0572-0137>.

33. Environmental Protection Agency (2020). Agency Information Collection Activities; Proposed Renewal of an Existing Collection (EPA ICR n..2517.03 and OMB Control N. 2070–0194); Comment Request. *Federal Register*, **18**. At:: <https://www.regulations.gov/document?D=EPA-HQ-OPPT-2010-0572-0195>.

34. National Nanotechnology Initiative (2016). Subcommittee on Nanoscale Science, Engineering, and Technology. Committee on Technology. National Science and Technology Council. The National Nanotechnology Initiative - Supplement to the President's 2017 Budget. Washington, pp. 7. At: <https://www.nano.gov/sites/default/files/pub_resource/ nni_fy17_budget_supplement.pdf>.

35. National Institute for Occupational Safety and Health (2019). Continuing to Protect the Nanotechnology Workforce: NIOSH Nanotechnology Research Plan for 2018 – 2025. By Hodson L, Geraci C, Schulte P. Cincinnati, OH: U.S. Department of Health and Human Services, Centers for Disease Control and Prevention, National Institute for Occupational Safety and Health. At <https://doi.org/10.26616/NIOSHPUB2019116>.

36. Jarvis, D. and Richmond, N. (2011). Regulation and governance of nanotechnology in China: Regulatory challenges and effectiveness. *European Journal of Law and Technology*, **3**.

37. Abbott, K. W, Marchant, G. E., and Sylvester, D. J. (2010). Transnational regulation of nano-technology: Reality or romanticism? In: Hodge, G. A., Bowman, D. M., and Maynard, A. D. (Eds.), *International Handbook on Regulating Nanotechnology*. Cheltenham (Edward Elgar), pp. 525–526.

38. Larsson, S., Jansson, M., and Boholm, Å. (2019). Expert stakeholders' perception of nanotechnology: Risk, benefit, knowledge, and regulation. *Journal of Nanoparticle Research*, **57**, 2019. https://doi.org/10.1007/s11051-019-4498-1.

39. Engelmann, W. (2016) Novos desafios para o Direito na era das nanotecnologias. *Tomo*, **29**, pp. 37–54.

40. Ramos, A. C. (2016) Direito internacional privado e o direito transnacional: entre a unificação e a anarquia. *Revista de Direito Internacional*, **2**, pp. 504–521.

41. Organization for Economic Co-operation and Development (2020). Testing Programme of Manufactured Nanomaterials – Sponsors. At: <http://www.oecd.org/chemicalsafety/nanosafety/sponsors-testing-programme-manufactured-nanomaterials.htm>.

42. Organization for Economic Co-operation and Development (2020). Publications in the Series on the Safety of Manufactured Nanomaterials. At: < http://www.oecd.org/env/ehs/nanosafety/publications-series-safety-manufactured-nanomaterials.htm>.

43. Organization for Economic Co-operation and Development (2018). OECD Nº 88: Investigating the different types of risk assessment of manufactured nanomaterials. Identifying tools available for risk management measures and uncertainties driving nano-specific data needs. Series on the Safety of Manufactured Nanomaterials. Paris.

44. International Organization for Standardization (2020). ISO Standards are Internationally Agreed by Experts. At: <https://www.iso.org/about-us.html?>.

45. ISO/IEC/NIST/OECD (2008). International workshop on documentary standards for measurements and characterisation in nanotechnologies. Final report, Gaithersburg, Maryland, USA, pp. 26-28. At: <https://www.pnnl.gov/nano/pdf/ISO_IEC_NIST_OECDwkshp.pdf>.

46. Mello, C. H. P., Silva, C. E. S., Turrioni, J. B., and Souza, L. G. M. (2009) ISO 9001:2008 – Sistema de Gestão da Qualidade para Operações de Produção e Serviços (Atlas, São Paulo).

47. International Organization for Standardization (2020). Standards By ISO/TC 229 Nanotechnologies. At: <https://www.iso.org/committee/381983/x/catalogue/>.

48. Engelmann, W. and Martins, P. S. (2017). A ISO, suas normas e estruturação: possíveis interfaces regulatórias. In: Engelmann, W. and Martins, P. S. (Orgs.) As Normas ISO e as Nanotecnologias: entre a autorregulação e o pluralismo jurídico (Karywa, São Leopoldo), pp. 79-80.

49. Horn, T. M. (2016). ISO 14001:2015 – Ciclo de Vida. Total Qualidade. At: <http://www.totalqualidade.com.br>.

50. Snir, R. and Ravid, G. (2015). Global nanotechnology regulatory governance from a network analysis perspective. Regulation & Governance. At: <http://www.ravid.org/papers/RG_2015.pdf>.

Chapter 11

Future: Green and Sustainable Nano

Gustavo Marques da Costa[a] and
Michele dos Santos Gomes da Rosa[b]

[a]*Instituto Federal de Educação Ciência e Tecnologia Farroupilha (IFFar)*
Campus Santo Augusto, Santo Augusto, CEP 98590-000, RS, Brazil
[b]*Cardiovascular Centre of Universidade de Lisboa–CCUL,*
Faculty of Medicine, Universidade de Lisboa, Portugal
markesdakosta@hotmail.com

The creation of artificial organs and implants with greater affinity for the original tissue with the use of biodegradable polymers or hydroxyapatite in biosynthetic films has been increasing in recent times. Nanotechnology can help in the use of surfaces containing nanometric grooves, or "stamped," even with adhesion molecules using, for example, the atomic force microscope, to improve cell adhesion and guide their growth in the desired shapes.

Thus, titanium alloy nanocomposites can be used to increase the longevity and biocompatibility of surgical devices and prostheses. Some examples: nerve cell implants grown in polymeric meshes for spinal cord repair; nerve cells grown in polymeric meshes for

Environmental, Ethical, and Economical Issues of Nanotechnology
Edited by Chaudhery Mustansar Hussain and Gustavo Marques da Costa
Copyright © 2022 Jenny Stanford Publishing Pte. Ltd.
ISBN 978-981-4877-76-3 (Hardcover), 978-1-003-26185-8 (eBook)
www.jennystanford.com

spinal cord repair; bone or cartilage cells for reconstitution of joints and liver cells for liver construction for transplantation. Miniature biological samples required in rapid diagnosis are being developed by microfluid and nano-techniques using particles such as quantum dots, gold nanoparticles, magnetic nanoparticles, or fullerene.

Ultrasensitive personal health monitors are expected to be available in a few years. Implantable devices in the body will be able to continuously monitor blood levels of certain biological indicators and automatically adjust the release of drugs in appropriate quantities. For real time, administer yourself the necessary doses of insulin.

Combined advances in genomics and nanotechnology, such as the use of nanopores to quickly measure the size of DNA molecules, should also result in the development of sensors that determine genetic constitution quickly and accurately, enabling knowledge of genetic predisposition to diseases.

However, the product life cycle must be skewed to ensure a safe and sustainable development of nanoproducts, being the great challenge to compile a combined view of the life cycle of a nanomaterial in relation to energy use and its toxicological impact. Most of the necessary data do not exist or are owned by industrial companies. Much more efforts and resources have to be put in the emerging field of green nanotechnology in order to enable a public dialogue based on facts and evidence-based research.

This chapter is the conclusion of the book seeing the future and the environment in a sustainable way that conducts to a future: green and sustainable nano.

11.1 Future

One of the impact promises associated with nanotechnology is that it will facilitate greener and more sustainable economic growth. Nanotechnology offers the promise of developing multifunctional materials that will contribute to building and maintaining lighter, safer, smarter, and more efficient vehicles, aircraft, spacecraft, and ships. In addition, nanotechnology offers various means to improve the transportation infrastructure (Paschoalino, Marcone, and Jardim 2010).

Research in the use nanotechnology for regenerative medicine spans several application areas, including bone and neural tissue engineering. For instance, novel materials can be engineered to mimic the crystal mineral structure of human bone or used as a restorative resin for dental applications. Researchers are looking for ways to grow complex tissues with the goal of one day growing human organs for transplant. Researchers are also studying ways to use graphene nanoribbons to help repair spinal cord injuries; preliminary research shows that neurons grow well on the conductive graphene surface (Paschoalino, Marcone, and Jardim 2010).

Nanomedicine researchers are looking at ways that nanotechnology can improve vaccines, including vaccine delivery without the use of needles. Researchers are also working to create a universal vaccine scaffold for the annual flu vaccine that would cover more strains and require fewer resources to develop each year (Pautler and Brenner 2010).

As discussed earlier, nano-engineered materials in automotive products include polymer nanocomposites structural parts; high-power rechargeable battery systems; thermoelectric materials for temperature control; lower rolling-resistance tires; high-efficiency/low-cost sensors and electronics; thin-film smart solar panels; and fuel additives and improved catalytic converters for cleaner exhaust and extended range (Farokhzad and Langer 2009; Hussain 2018; Hussain 2020a; Hussain 2020b).

Nano-engineering of aluminum, steel, asphalt, concrete, and other cementitious materials and their recycled forms offers great promise in terms of improving the performance, resiliency, and longevity of highway and transportation infrastructure components while reducing their life cycle cost. New systems may incorporate innovative capabilities into traditional infrastructure materials, such as self-repairing structures or the ability to generate or transmit energy (Costa and Jamami 2001).

Nanoscale sensors and devices may provide cost-effective continuous monitoring of the structural integrity and performance of bridges, tunnels, rails, parking structures, and pavements over time. Nanoscale sensors, communications devices, and other innovations enabled by nanoelectronics can also support an enhanced transportation infrastructure that can communicate with vehicle-based systems to help drivers maintain lane position, avoid collisions, adjust travel routes to avoid congestion, and improve drivers' interfaces to onboard electronics (Barua, Datta, and Das 2020).

"Game changing" benefits from the use of nanotechnology-enabled lightweight, high-strength materials would apply to almost any transportation vehicle. For example, it has been estimated that reducing the weight of a commercial jet aircraft by 20% percent could reduce its fuel consumption by as much as 15%. A preliminary analysis performed for NASA has indicated that the development and use of advanced nanomaterials with twice the strength of conventional composites would reduce the gross weight of a launch vehicle by as much as 63%. Not only could this save a significant amount of energy needed to launch spacecraft into orbit, but it would also enable the development of single stage to orbit launch vehicles, further reducing launch costs, increasing mission reliability, and opening the door to alternative propulsion concepts (Barua, Datta, and Das 2020).

The use of non-terrestrial materials is generally viewed as necessary for the development of space-based infrastructure, because of the cost of importing material out of Earth's gravity well. In addition, non-terrestrial materials are increasingly proposed for terrestrial use, a suggestion motivated originally by the "doomsday" scenarios of the early 1970s, which purported to demonstrate that shortages of energy and materials would shortly cause the collapse of industrial civilization (Costa and Jamami 2001).

Geochemically rare but important metals have been a particular focus of concern, especially since with present technology they must be extracted from rare, sporadic, and anomalous deposits ("ore bodies") where geologic happenstance has concentrated them. As several such metals are siderophile (e.g., Ni, Co, precious metals), sideritic asteroidal bodies have seemed especially promising as non-terrestrial sources for them, even for terrestrial use. The fact that these metals are either strategic (e.g., Ni, Co), precious (Au), or both (Pt) has made them particularly attractive (Adlakha-Hutcheon et al. 2009).

No shortages of metals have appeared in the intervening 25 years since the Limits to Growth appeared; indeed, the general trend of metals prices has been downward, and this has been reflected in a severe downturn of (for example) the job prospects of geologists specializing in ore deposits. Nonetheless, this is widely perceived as a temporary trend fueled by imports from developing countries, where abundant high-grade ores are still available. At such point as these ores begin to dwindle, especially as domestic demand in the developing world increases, it is commonly felt that the Limits to Growth scenarios will return with a vengeance (Bawa 2011).

However, new technological developments are calling the assumptions behind these scenarios into question. In particular, the embryonic field of "molecular nanotechnology," in which macroscopic objects are designed and constructed at atomic scales, promises to invalidate the implicit assumptions on which the traditional paradigms of extractive metallurgy are based.

Indeed, I will argue that pollution control will be a major economic driver for the development of these technologies, with one result being that the distinction between a "resource" and a "pollutant" is likely to become blurred. Finally, I will examine the implications for space resources from this new perspective (Braga et al. 2001).

11.1.1 The Application

Air purification with ions, wastewater purification with nanobubbles, or nanofiltration systems for heavy metals are some of its environmentally friendly applications. Nanocatalysts are also available to make chemical reactions more efficient and less polluting (Cançado et al. 2006; Hussain 2020c).

Thus, although energetically cheap element separation at the molecular level is conceptually possible, it seems at first sight a long way off because of the enormous R&D effort needed to develop a full MNT capability. Drexlerian molecular assemblers are not imminent, and unless developmental pathway(s) can be identified, the relevance of MNT to space development over the near to mid-term is unclear (Brook et al. 2010).

It seems such pathways exist, however. First, it appears an interim middle ground exists between present-day "shake and bake" synthesis approaches and full-scale Drexlerian assemblers. One such pathway is molecular self-assembly, which is receiving much study, and which conceptually provides a way of atomic-scale structuring without assemblers. Another way is to use primitive assemblers, such as gangs of scanning-probe microscopes themselves probably made by conventional microlithography techniques, to "sculpt" molecularly perfect structures on surfaces. Although the number of atoms that can be individually arranged this way remains minuscule, it may be practical for constructing highly selective, tailored catalytic surfaces. In this way, the intrinsic parallelism of solution chemistry can be exploited; synthesis yields could be vastly improved through effectively excluding the non-catalyzed reaction pathways. A major problem in conventional chemical synthesis is the number of unwanted by-products that form because the synthesis reactions are too unselective; this not only decreases yield but leads to separation and disposal problems. Furthermore, the decrease in yield is substantial for syntheses requiring multiple steps. (Again, this approach is anticipated by biosystems, which use highly specific catalysts—enzymes—to direct particular synthesis pathways in living cells) (Boisseau and Loubaton 2011; Costa and Jamami 2001).

In addition, such "nanostructured" materials have the advantage of no moving parts, which eases the engineering difficulties. Nonetheless, the developmental problems remain formidable, and financial incentives must exist if they are to be developed over relevant timescales (Organization et al. 2017).

Such an economic driver exists: pollution control. As noted already, it is merely another aspect of separating atoms. Thermal-based approaches, however, as used in traditional pyrometallurgy, are obviously impractical because of the low concentrations involved. Although isothermal phase changes can be used (e.g., precipitation), they still have serious limitations; they require additional reagents (which probably were purified by pyrometallurgy), and there is little control over the precipitates as their nature is set by the laws of chemistry. To wit, there are definite limits (set by the solubility products) to the concentrations that can be treated; the nature of the precipitated phase may be inconvenient (e.g., through being vulnerable to oxidation); and finally, changes in solution composition can cause unwanted phases to form, depending on the species present, their concentrations, and the stability fields of possible solid phases (Costa and Jamami 2001).

Selectivity is also an issue; it is common to have low levels of a toxic ion (e.g., Pb) among a much larger concentration of an innocuous ion (e.g., Ca), and a practical extractive process thus must strongly discriminate in favor of the rare ion.

More promising approaches to separation involve nanostructured materials, such as highly selective semipermeable membranes, which could filter out and concentrate particular solutes (e.g., heavy metals). Already, molecular sieves such as zeolites are used to separate gaseous N_2 from O_2, but wider use of such separation is hindered by the expense of crystallizing the sieves. Specific adsorbers, with molecular binding sites highly specific for certain ligands, furnish another example. Note also that such devices do not involve nanotechnological machines, i.e., devices with molecular-scale moving parts; they instead operate passively (Farokhzad and Langer 2009).

Both selectivity and extraction of solutes at low concentration require precision at atomic scales; indeed, current limitations in the applications of membrane technology largely result from fabrication difficulties. The materials are expensive, rather delicate, and molecularly imprecise. Hence, the desirability of atomically precise assembly provides incentives for near-term nanofabrication techniques (Chan 2006).

In addition, a vast and growing literature exists on highly specific complexing agents (typically macrocyclic compounds such crown ethers and calixarenes) for various metal ions, both for potential therapeutic uses as well as for environmental and hydrometallurgical applications. However, many such compounds are not currently economic due to their costly syntheses. Hence, directed catalysis, as by nanotailored catalytic surfaces as described earlier, may make such compounds economic and provide another economic motivation for "interim" nanotechnology (Barua, Datta, and Das 2020).

Initially, pollution control will drive these technologies because it is the high value application; the value of the extracted material itself will be insufficient to pay for the technology. Applications initially will lie in such areas as the clean-up of industrial wastewater streams, which is required before their discharge into surface waters. Environmental remediation, such as the clean-up of dump and mining sites (in particular, the amelioration of acid drainage resulting from the oxidation of reactive tailings), is also an obvious near-term application (Farokhzad and Langer 2009; Wallington, Sullivan, and Hurley 2008).

As these technologies mature and their costs fall, however, they will ultimately blur the distinction between a "pollutant" and a "resource"; that is, the value of the extracted material will become important in itself. Indeed, as demand increases, many sources containing metals in aqueous solution may become attractive. (Note also that the by-product of such extraction processes would be pure water, which hardly poses a disposal problem.) Seawater is an obvious possibility, but highly saline natural brines may be more attractive. Indeed, deep, saline groundwaters such as those associated with oilfields currently pose a disposal problem (Brook et al. 2010).

Because these technologies involve extraction from solutions, it may also prove economic to leach materials containing useful elements and recover metals from the leachates. This could be looked on as an extension of present hydrometallurgy, as with the present-day cyanide-based solution extraction of Au, or the leaching of Cu with dilute H_2SO_4. For a further example, during World War II an experimental process for magnesium production involved the dissolution of olivine (($Mg,Fe)SiO_4$) by a strong mineral acid, such as HCl. Obviously, Fe could be a by-product (unwanted at the time); these authors also noted that Ni and Co, which commonly substitute for Fe and Mg at concentrations up to 2000 and 130 ppm, respectively, could be recovered. Finally, another unwanted by-product was silica gel formed from the disaggregated mineral, but this may itself prove useful in a silicate-based nanotechnology, as discussed below. Olivine is a ubiquitous mineral; it makes up most of planetary mantles and is locally abundant at the surface of both the Earth and the Moon.

One hindrance to wider application of such solution-based extractive processes has been the necessity for selective extraction of solutes from dilute solutions. This is the very problem of pollution control again and underscores the fundamental fact that what's a "resource" and what's a "pollutant" is merely a matter of perspective. Note also that biosystems have anticipated a solution-based approach to extracting raw materials; consider digestion (Barua, Datta, and Das 2020; Hou et al. 2019).

In addition, over the longer term, MNT is likely to change substantially what elements are desired. In particular, current technology relies heavily on metals for structural members. It is well known, however, that ordinary macroscopic materials are a couple of orders of magnitude weaker than the ultimate strength limits set by chemical bonds because of their extremely high densities of defects, such that the strengths are determined instead by such things as grain boundaries and dislocations. Under such circumstances, metals are useful because they are highly tolerant of microflaws, even at extreme densities; incipient cracks tend to "heal" via plastic deformation rather than propagate. However, metals are intrinsically weak because of this readiness to deform; brittle materials are potentially far stronger, but liable to catastrophic failure via propagation of Griffith cracks unless they are essentially defect free at a molecular level. MNT should allow fabricating such defect-free materials, with profound potential consequences.

Most theoretical studies have focused on tetrahedral (sp3) carbon frameworks ("diamondoid") as the structural basis of MNT. This is partly motivated by the enormous strength/weight ratio theoretically possible with such networks, but the familiarity and vast knowledge base of organic chemistry also provide a motivation. However, silicates, compounds of Si and O, are a potentially valuable alternative. Silicates are based on an SiO_4 tetrahedron that easily enters 3D coordination; that is, each vertex can be shared with an adjacent tetrahedron such that all oxygens "bridge" between two silicon atoms. Furthermore, the Si–O bond is strong and directional, due to its partial covalent character. Moreover, in contrast to "diamondoid" carbon, silicates can polymerize at STP, even from aqueous solution; hence a silicate-based MNT may well be nearer term (Coppola 2003).

Finally, the crust of the Earth is largely made of silicates; oxygen and silicon, respectively, make up 60.4 and 20.5 atom percent of the crust, and thus raw materials are literally everywhere. However, conventional ores are seldom silicates, simply because of the difficulty in breaking up the Si–O bonds with current pyrometallurgy. Indeed, the waste from conventional mining largely consists of comminuted silicates; ore minerals are typically sulfides and must be separated from the silicate "gangue" by grinding and flotation. The left-over silicate debris ("tailings") currently constitutes an environmental problem; it is unesthetic, commonly constitutes a dust hazard, and the oxidation of residual sulfides commonly leads to acidic drainage. Its very comminution, however, suggests that tailings might be ideal feedstock for a silicate-based MNT, and certainly there would be no environmental objection to its reprocessing (Bosetti and Vereeck 2011).

Ironically, therefore, the silicates that make up the bulk of the Earth, and that have been ignored in traditional resource scenarios, may yet prove to be among the most valuable raw materials for a truly mature technology. Indeed, the metals such as Fe, Al, Mg, and so on that make up a large percentage of common rocks may ultimately become a (largely) unwanted by-product of a silicate-based nanotechnology.

Calling MNT "convenient" for space applications is likely to be a major understatement; it may indeed be vital for a viable off-Earth civilization. The value of the extreme strengths of MNT-based materials is only one aspect; as described earlier for terrestrial uses, MNT also makes practical a wide variety of raw materials. I had previously argued, based on several millennia of terrestrial experience, that anomalous concentrations of elements—"ores"—would be necessary for space-based resource extraction, just as they have been on Earth. With the advent of MNT, this seems merely another naive extrapolation of current technology. In addition, because structural metals are likely to be unimportant even with a relatively immature MNT, the ready availability of even high-quality Ni-alloy steels on sideritic asteroids may prove irrelevant (Globus et al. 1998).

Of course, C is also abundant in carbonaceous chondrite-like bodies, and thus asteroidal bodies may still prove to be extremely attractive sources of raw materials, quite apart from their low gravity wells. Conversely, C is nearly absent from many rocky solar system bodies, the Moon in particular, so a diamondoid-based MNT seems unattractive there. (Parenthetically, however, it might be noted that the largest off-Earth reservoir of C in the inner solar system is the CO_2 atmosphere of Venus, which thus may have unexpected long-term value) (Packwood 2020).

However, a silicate-based nanotechnology is likely to find many applications in space, as silicates dominate rocky bodies such as the Moon just as they do the Earth. Indeed, the regoliths mantling bodies like the Moon, which consist of silicate debris comminuted by eons of meteoritic impact, may prove to be ideal feedstocks. A silicate MNT devised to handle terrestrial mining debris should be readily adaptable to such regoliths (Globus et al. 1998).

11.2   Conclusion

The separation of elements at an atomic level is an obvious near-term application of molecular nanotechnology. Viewed in one way, this is the problem of resource extraction; but viewed in another, it is the problem of pollution control. Indeed, pollution control is likely to be an economic driver for molecularly precise fabrication, because of the ongoing financial incentives involved.

This has two major implications for space development. First, materials from lunar or asteroidal mines are unlikely to be significant for terrestrial use; when desired elements can be recovered at ppm levels from aqueous solutions, whether wastewater streams, leachates, or natural brines, bringing them in from space is unlikely to make economic sense. Moreover, under a "total product lifetime closure" approach, even space-derived material will have hidden environmental costs due to its ultimate costs of disposal, and such costs will have to be addressed in any case.

Second, by the same token, such technologies vastly broaden the potential sources of raw material in space for development in space. When even low concentrations of a desired element can be exploited, "ores" in the traditional sense become unnecessary. However, MNT is also likely to change considerably the desired elements; in particular, structural metal is likely to become unimportant, whereas carbon will become highly sought after. More unexpectedly, the silicates that make up the bulk of the rocky bodies in the inner solar system may also prove extremely valuable for MNT applications. Hence, the comminuted, rocky regoliths of bodies such as the Moon may prove to be ideal feedstocks, especially as a silicate nanotechnology is likely to be developed in any case for terrestrial applications.

References

Adlakha-Hutcheon, G, R Khaydarov, R Korenstein, R Varma, A Vaseashta, H Stamm, and M Abdel-Mottaleb. 2009. "Nanomaterials, Nanotechnology." In *Nanomaterials: Risks and Benefits*, 195–207. Springer.

Barua, Ranjit, Sudipto Datta, and Jonali Das. 2020. "Application of Nanotechnology in Global Issues." In *Global Issues and Innovative Solutions in Healthcare, Culture, and the Environment*, 292–300. IGI Global.

Bawa, Raj. 2011. "Regulating Nanomedicine-Can the FDA Handle It?" *Current Drug Delivery* **8** (3): 227–34.

Boisseau, Patrick, and Bertrand Loubaton. 2011. "Nanomedicine, Nanotechnology in Medicine." *Comptes Rendus Physique* **12** (7): 620–36.

Bosetti, Rita, and Lode Vereeck. 2011. "Future of Nanomedicine: Obstacles and Remedies." *Nanomedicine* **6** (4): 747–55.

Braga, Alfesio, Luiz Alberto Amador Pereira, György Miklós Böhm, and Paulo Saldiva. 2001. "Poluição Atmosférica e Saúde Humana." *Revista USP*, no. **51**: 58–71.

Brook, Robert D, Sanjay Rajagopalan, C Arden Pope III, Jeffrey R Brook, Aruni Bhatnagar, Ana V Diez-Roux, Fernando Holguin, et al. 2010. "Particulate Matter Air Pollution and Cardiovascular Disease: An Update to the Scientific Statement from the American Heart Association." *Circulation* **121** (21): 2331–78.

Cançado, José Eduardo Delfini, Alfesio Braga, Luiz Alberto Amador Pereira, Marcos Abdo Arbex, Paulo Hilário Nascimento Saldiva, and Ubiratan de Paula Santos. 2006. "Repercussões Clínicas Da Exposição à Poluição Atmosférica." *Jornal Brasileiro de Pneumologia* **32**: S5–11.

Chan, Vivian SW. 2006. "Nanomedicine: An Unresolved Regulatory Issue." *Regulatory Toxicology and Pharmacology* **46** (3): 218–24.

Coppola, D. 2003. "Nanocrystal Technology Targets Poorly Water-Soluble Drugs." *Pharm Tech* **27**: 20.

Costa, Dirceu, and Mauricio Jamami. 2001. "Bases Fundamentais Da Espirometria." *Rev Bras Fisioter* **5** (2): 95–102.

Farokhzad, Omid C, and Robert Langer. 2009. "Impact of Nanotechnology on Drug Delivery." *ACS Nano* **3** (1): 16–20.

Globus, Al, David Bailey, Jie Han, Richard Jaffe, Creon Levit, Ralph Merkle, and Deepak Srivastava. 1998. "Nasa Applications of Molecular Nanotechnology."

Hou, Xuyan, Yilin Su, Shengyuan Jiang, Pan Cao, Pingping Xue, Tianfeng Tang, Long Li, and Tao Chen. 2019. "SPACE CLIMBING ROBOT FEET WITH MICROARRAY STRUCTURE BASED ON DISCRETE ELEMENT METHOD." *International Journal of Robotics and Automation* **34** (1).

Hussain, Chaudhery Mustansar. 2018. *Handbook of Nanomaterials for Industrial Applications*. Elsevier.

———. 2020a. *Handbook of Functionalized Nanomaterials for Industrial Applications*. Elsevier.

———. 2020b. *Handbook of Manufacturing Applications of Nanomaterials*. Elsevier.

———. 2020c. *The ELSI Handbook of Nanotechnology: Risk, Safety, ELSI and Commercialization*. John Wiley & Sons.

Organization, World Health et al. 2017. *WHO Guidelines on Protecting Workers from Potential Risks of Manufactured Nanomaterials*. World Health Organization.

Packwood, Daniel M. 2020. "Exploring the Configuration Spaces of Surface Materials Using Time-Dependent Diffraction Patterns and Unsupervised Learning." *Scientific Reports* **10** (1): 1–11.

Paschoalino, Matheus P, Glauciene PS Marcone, and Wilson F Jardim. 2010. "Os Nanomateriais e a Questão Ambiental." *Química Nova* **33** (2): 421–30.

Pautler, Michelle, and Sara Brenner. 2010. "Nanomedicine: Promises and Challenges for the Future of Public Health." *International Journal of Nanomedicine* **5**: 803.

Wallington, Timothy J, John L Sullivan, and Michael D Hurley. 2008. "Emissions of Co2, CO, NOx, HC, PM, HFC-134a, N2o and Ch4 from the Global Light Duty Vehicle Fleet." *Meteorologische Zeitschrift* **17** (2): 109–16.

Index

absorption 10, 64, 66–68, 156
acceptance 110, 115, 116, 119,
    121, 127, 143
  industrial 181
  public 118, 119, 122
  social 110, 114, 115, 132
actors 142, 143, 178, 185, 192,
    198
additive 41, 115
  fuel 207
adhesion 14, 20, 189
adsorption 6, 38, 48, 79, 80, 82
adverse effect 5, 14, 63, 66, 102,
    152, 168
AFM *see* atomic force microscope
agglomeration 6, 79, 93, 198
aggregation 79–81, 83, 93, 138
agriculture 77, 111, 131, 136, 153,
    160, 167, 178
air filter 8, 14, 39
  functional 22
  nanofibrous 20
air filtration 8, 14, 15, 17, 20–22,
    32, 34, 37, 39–41, 48, 49
airflow 16–18, 21, 100
air quality 101, 103, 111
air velocity 18, 19, 39
airways 68, 94, 96, 100, 101
alveoli 62, 93, 95, 103
anthropogenic origin 3, 13, 63, 75,
    78, 91
application 1, 2, 4, 5, 7, 22, 32, 40,
    45–47, 49, 74–76, 115, 116,
    126, 127, 166, 168, 169, 178,
    179, 191, 198, 207, 211, 212
  air-filtration 19, 20, 35
  biological 40
  drug-delivery 36

electrochemical 41
filter 7
food-related 116
hydrometallurgical 211
liquid microfiltration 40
nanotechnology-enabled 167
pharmaceutical 38
photochromic 41
photo-sensing 46
piezoelectric 40
sensor 167
technological 2
terrestrial 215
therapeutic 67
tissue regeneration 44
assessment 12, 13, 64, 65, 98, 111,
    113, 123, 136, 179, 184, 185,
    188, 197
  environmental 102
  toxicological 185
asthma 13, 68, 93, 94, 99, 101
atomic force microscope (AFM) 2,
    9, 10, 33, 47, 205
attitude 115, 117, 121, 127, 128,
    142
autonomy 118, 119, 125

bacteria 3, 8, 14, 35, 154
battery 8, 31, 38, 44, 154
bioaccumulation 12, 65, 155, 198
bioavailability 74, 76, 78–82, 84,
    104
bioethics 133, 134, 138
biomedicine 7, 31, 153
biomolecule 62, 63, 169
biosensor 42, 165
biosystem 62, 63, 83, 210, 212
bloodstream 62, 68, 90, 93, 94

Boltzmann distribution 18
brain 64, 66, 68, 154
bronchioles 92, 95
Brownian diffusion 16
Brownian motion 11, 17

carbon nanotube 7, 63, 65, 74, 91,
    94, 153
catalysis 7, 31, 43, 45, 211
catalyst 6, 8, 36, 48, 210
cell 61–64, 66, 67, 83, 118, 170,
    210
  cartilage 206
  diseased 170
  flattened epithelial 95
  healthy 170
  liver 206
  nerve 205
  phagocytic 68
  red blood 133
  stem 118
challenges 133, 135, 137, 140,
    146, 172, 173
  global 194
  medical 170, 171
  societal 186
  socio-environmental 192
characterization 1, 8–11, 46, 47,
    65, 69, 75, 147, 155, 157, 159,
    168, 169
  chemical 160
  physical 32
  physicochemical 185
characterization technique 9, 11,
    46–49
chemical 160, 179, 184, 188, 191,
    193
  carcinogenic 93
  conventional 160
chemical substances 66, 69, 102,
    184, 187, 188
coating 36, 82, 83, 166
coherent ethico-legal plan
    141–143, 145, 146

coherent life plan 132, 135,
    140–142, 145
collaboration 142, 159, 167, 182,
    193
collection efficiency 17–21, 32, 35,
    39, 40
commercialization 65, 109, 122,
    124, 138, 146, 152, 188
composites 37, 40, 208
conception 139, 141, 142, 145,
    180
consumer 115–117, 119,
    125–127, 155, 179, 184
consumer product 63, 65, 69, 79,
    156, 160
contamination 14, 75, 155, 160,
    161, 189
cosmetics 12, 67, 113, 136, 166,
    178, 184, 187
crystalline structure 5, 6, 47, 78,
    198

damage 13, 22, 66, 83, 90, 91, 103,
    133, 139, 155, 156, 170
  environmental 159
  lung 91, 166
  oxidative 69
  oxidative DNA 66
deposition 15–17, 20, 22, 64, 75,
    79, 93, 104
device 2, 14, 65, 120, 172, 173,
    207, 211
  communications 207
  electronic 40, 77
  gallium nitride radio frequency
    172
  high-frequency 172
  medical 120, 170, 187
  microscopic recording 166
  nano-computation 113
  nanostructured 8
  nanotechnology-based 117
  photovoltaic 32
  surgical 205

diagnosis 63, 166, 169, 170, 174, 206
diffusion 5, 17, 20, 81, 82, 116
discovery 110, 116, 133, 135, 172
disease 68, 69, 92, 94, 101, 102, 105, 161, 167, 170, 172, 206
  Alzheimer's 68, 170, 171
  cardiorespiratory 102
  cardiovascular 103
  chronic lung 92, 101
  functional obstructive 98
  genetic 120
  heart 13, 98
  infectious 14, 170, 171
  neurodegenerative 68
  obstructive pulmonary 101
  organic obstructive 98
  Parkinson's 68, 170, 171
  respiratory 98, 100, 102–104
  systemic 102
  viral 94
dissolved organic matter (DOM) 81, 82
DLS *see* dynamic light scattering
DOM *see* dissolved organic matter
drug 67, 153, 154, 159, 166–170, 173, 206
  anti-tumor 153
  chemotherapy 170
  lipophilic 104
  low-soluble 168, 169
dust 15, 90, 92, 93
dynamic light scattering (DLS) 9, 11

economic driver 209, 210, 215
economy 114, 122, 153, 161, 167, 186, 192
elasticity 20, 38, 95
electrospinning 16, 21, 22, 32, 34–36, 42, 43, 48
  coaxial 36
  conventional 40

ELSI *see* ethical, legal, and social implications
endocytosis 62, 67, 104
energy 2, 6–8, 47, 153, 161, 165, 166, 208
  nuclear 137
  polluting 159
  sustainable 153
  vibrational 10
environment 2, 4, 6–8, 10–14, 22, 61, 62, 64, 65, 69, 73–75, 78–81, 116, 117, 132–134, 138, 139, 143, 144, 152, 153, 155, 156, 160, 161, 177–180, 183–188, 190–193
environmental impact 111, 113, 127, 134, 173
environmental issues 7, 14, 152, 157, 173, 180
epithelium
  alveolar 103
  pulmonary 67, 94
  respiratory 104
ethical, legal, and social implications (ELSI) 132, 135, 138, 144, 185
ethical acceptability 110, 126, 132, 135, 145
ethical issues 110, 112, 117, 118, 123, 125, 128, 133, 145
ethics 109, 121, 131, 132, 134, 143–145, 179, 180, 198
evaluation 97, 98, 102, 109, 111, 113, 160, 184
exposure assessment 100, 156, 160, 185

fabrication 33, 46, 113, 124, 165, 211, 215
fiber 14–20, 33, 36–40, 46, 47
  cylindrical 17
  glass 39
  lyocell 41
  micrometric 33

microscale 21
micro-sized 20
mineral 93
polypropylene 40
polyurethane 36
shell 36
submicrometric 45
fibrosis 68, 94, 105
filter media 15, 16, 18–21, 32, 35,
    39, 46, 48, 49
    commercial 39
    fibrous 15
    fibrous air 14
    nanofiber 32
filtration 15, 16, 18, 20, 21
fluid 7, 16, 34, 36
food production 4, 111, 117, 153
food supply 78, 119
force 10, 16, 22, 32, 37, 95, 114,
    184
    adhesion 15
    drag 21
    electrical 22
    electrostatic 37
    hydration 80
    intermolecular 45
    viscous 34
Fourth Industrial Revolution 131,
    132, 138
fullerene 6, 7, 206

gaps 5, 41, 112, 133, 161, 182,
    189, 194
gel 22, 32, 41, 44, 48, 154, 172,
    212
genome 64, 66
graphite 7, 91, 154

health 22, 92, 93, 103, 111, 113,
    144, 145, 153–159, 166, 169,
    178–181, 188, 189, 194, 195,
    197
    adverse 195
    environmental 63, 64, 153, 157

human 4, 6, 62, 64, 66, 68–70,
    101, 102, 132, 134, 156, 160,
    161, 183–188, 191–193
    nanotechnology for 167, 173
    people's 102
    public 13, 14, 159, 161, 179,
        187
    worker's 190
health risk 69, 90, 93, 103, 125
human being 12, 14, 69, 103, 124,
    138, 142, 155, 157, 198
human life 12, 89, 116, 133, 135,
    137, 139

industry 109, 111, 112, 115, 119,
    121, 152, 153, 159–161, 168,
    171, 173, 182, 184, 186
    automotive 131
    ceramic 64
    pharmaceutical 165
inflammation 92, 94, 101
initiative 142, 158, 179, 185, 186,
    193, 197
innovation 119, 126, 132, 133,
    138, 139, 143–145, 157, 161,
    167, 179, 183–186, 188, 194,
    198
inspiration 90, 95–97, 99
international trade 182, 196, 198
investment 112, 118, 152, 153,
    170, 177, 180, 186
ion 11, 79, 209, 211
ionic strength 74, 78, 81, 82

judgment 109, 140
    informed 122
    normative 110

law 2, 8, 109, 133, 134, 138–140,
    146, 147, 155, 158, 178, 179,
    191, 210
    international 158
    model 192
    natural 139, 140

legislation 14, 146, 179, 182, 187, 195
life cycle 12, 65, 134, 138, 143, 145, 206
liposomes 63, 104, 168
liquid filtration 34, 38, 39, 45, 46, 48
lung cancer 13, 68, 92
lung capacity 93, 96–98, 100, 101

macrophages 64, 66, 68, 104
material 2–9, 11, 12, 15, 21, 22, 33, 34, 37, 38, 62, 64–69, 73–75, 79, 80, 82, 152, 154, 156, 166, 167
  biological 10, 65, 118
  brittle 213
  cementitious 207
  colloidal 80
  conventional 6
  defect-free 213
  high-strength 208
  humic 79
  inorganic 11
  molten 33
  nanoceramic 64
  nano-engineered 207
  particulate 8, 15, 111
  polymeric 6, 42
  space-derived 215
  synthetic 45
  thermoelectric 207
mechanism 16, 17, 64, 67, 68, 78, 79, 94, 102, 167, 192
  chemical defense 66
  diffusional 18
  transfer 79
media 5, 15, 21, 79, 114, 127, 190
  biological 83, 168, 169, 174
  communications 115
  filtering 15
  membrane/filter 48
  nanoporous 173

medicine 7, 32, 69, 89, 110, 113, 120, 136, 166, 167, 169, 170, 174
  preventive 171
  regenerative 207
membrane 14, 15, 22, 32, 35, 41, 45, 46, 48, 49, 67, 95
  anodic aluminum oxide 41, 42
  cell 61, 68, 83
  ceramic 42
  electrospun 35
  fibrous 15
  hollow fiber 40
  lung 90
  mucous 61, 83
  nanoporous 41
  nanotechnological 154
  polycarbonate 41
  polymeric 15
  polystyrene 21
  semipermeable 211
  ultrafiltration 46
  water filtration 154
metal 7, 11, 34, 40, 78, 80, 92, 208, 212–214
  absorbing 7
  heavy 209, 211
  inhaled 92
  precious 208
  structural 214, 215
  transition 91
metamorphosis 137–139
microorganisms 9, 11, 13, 14, 159
microparticles 16, 168
model 82, 132, 140, 141, 168
  autocratic 125
  democratic 125
  ethico-normative 140
  information-plus-debate 125
  operant 124
  output 124
  radical positivist 134
  thermodynamic 82
  traditional regulatory 191

muscle 95, 96
  abdominal 96
  expiratory 96
  intercostal 95, 96
  respiratory 96

nanocosmetics 184, 185
nanoethics 110, 132, 143, 145
nanofiber 3, 6, 7, 15, 16, 18,
    20–22, 31–39, 41, 43–46, 49,
    195
  carbon 7
  cellulose acetate 36
  core–shell 7, 36
  electrospinning polyimide 20
  electrospun 14, 36
  hexadecane 36
  polyimide 21
  polymer 34
  polymeric 8, 22
  polypropylene 39
  semiconducting 3
nanomaterials 1–12, 22, 61–70,
    74, 75, 77, 90, 91, 93, 94, 143,
    144, 155–161, 168, 169, 172,
    173, 179, 180, 182–187,
    189–195, 197, 198
  biological 7
  carbon-based 7
  clothing-containing 74
  engineered 74, 134, 145, 181
  magnetic 196
  manufactured 4, 12, 69, 83, 147,
    181, 192–195
  safety of 179, 185, 192, 193
  synthetic 152
  toxicity of 6, 82, 83, 196
nanomedicine 169–173
nanoparticles (NPs) 3, 4, 6, 12,
    13, 22, 40–42, 62–69, 73–75,
    77–83, 89–94, 100–104, 155,
    159–161, 168–170, 188, 189,
    194, 195
  biodegradable 63
  crystalline 3
  filter 16
  gold 206
  inhalation of 62, 93
  inhaled 68
  interaction of 66, 68
  magnetic 206
  metal 12, 34
  naphthopyran-functionalized
    silica 41
  non-engineered 74
  silver 9, 11, 36, 42, 154
  soil 78
  toxicity of 82, 83
nanoproducts 65, 136, 152, 177,
    190, 192, 206
nanoscale materials 2, 5, 160, 166
nanoscience 110, 112–116, 118,
    120, 122, 126, 128, 132, 134,
    135, 165, 167, 180, 181, 183
nano-security 64, 156, 184, 198
nanostructure 6, 8, 32, 33, 35, 37,
    39, 41, 43–46, 49, 190
  boehmite 46
  complex 173
  metallic 45
  semiconductor 151
nanotechnology research 119,
    132, 135, 167, 190
nanotoxicology 61, 63, 83
norms 117, 183, 188
  ethical 116
  technical 194
NPs *see* nanoparticles

ores 209, 213–215
organism 5, 64, 67, 83, 93
  cellular 79
  living 7, 11, 63, 74, 83, 155, 156,
    160
organ 64, 66, 67, 93–95, 103, 120,
    155
  artificial 205
  spongy 95

oxidation 8, 78, 210, 212, 214

particulate matter 8, 13, 73, 101
pathway 93, 104, 209
  developmental 209
  heat rejection 173
  non-catalyzed reaction 210
  non-invasive 104
  potential exposure 93
  synthesis 210
  thermal 173
patient 98, 105, 120, 154, 168
PEC *see* predicted environmental
  concentration
penetration 13, 15, 36, 93, 159,
  171
permeability 20, 21, 40, 61, 67, 83
personal protective equipment
  (PPE) 35, 90, 103
perspective 13, 22, 112, 137, 145,
  146, 172, 212
  ethico-legal 138
  global 180
  mechanistic 101
  transdisciplinary 134
phase separation 22, 32, 44–46,
  48
phenomena 21, 34, 112, 113, 153,
  178, 197
pneumonitis 92
political system 125, 126, 198
pollutants 8, 79, 91, 93, 94, 96,
  101, 209, 212
  air 13, 94, 101
  atmospheric 13
  chemical 94
  coexisting 79
  inorganic 75
  poisonous 13
pollution 8, 14, 111
  air 1, 8, 22, 94, 101, 103
  environmental 13, 102
  indoor 91
  industrial 91

marine 91
polymer 11, 20, 34, 36–40, 44, 91,
  168
  biodegradable 205
  melted 38, 47
  non-thermoplastic 40
  synthetic 35
polystyrene 20, 37, 91
polyurethane 37, 40
population 11–13, 65, 102, 103,
  116, 117, 120, 121, 123, 126,
  158, 170, 185
porosity 14, 18–20, 22, 44, 46, 104
PPE *see* personal protective
  equipment
predicted environmental
  concentration (PEC) 75, 77
pressure drop 15, 16, 21, 32, 39
principles 2, 9, 118, 120, 125, 132,
  134, 135, 138, 144, 145, 152,
  179, 183, 188
  bioethical 134
  environmental 196
  ethical 118, 179
  general 132, 134, 143, 144, 183
  precautionary 69, 136, 144,
    161, 184
product 12, 65, 74, 75, 90, 115,
  116, 119, 136, 152, 155–160,
  184, 187–189, 191, 193,
  196–198
  aesthetic 153
  automotive 207
  commercial 39, 82
  diagnostic 171
  electronic 187
  innovative 168
  nano-enabled 147
  nanomedical 180
  tobacco 187
  veterinary 187
protein 11, 62, 63, 66, 79, 104, 168
  animal 45
  whey 68

protocol 156, 161, 162, 190

quality factor 20, 21, 35
quantum dot 3, 6, 8, 74, 206

radiation 10, 82, 187
Raman scattering 10, 11
raw material 44, 46, 49, 123, 152,
  212–215
reactive oxygen species (ROS) 64,
  66, 82
redox 74, 79–81
regulatory framework 145, 158,
  161, 178, 179, 187, 191, 198
regulatory initiative 177, 178, 186,
  187, 189, 192, 197, 198
research integrity 132, 135, 144,
  183
respiratory system 13, 62, 93,
  95–97, 99, 103
respiratory tract 67, 94, 104
rights
  fundamental 132–134, 144, 183
  intellectual property 155
  unprecedented 137
risk 11, 12, 64–70, 102, 103, 118,
  125–127, 132–137, 139,
  141–143, 146, 155–157, 159,
  180, 181, 185–189, 191, 197
  decision 139
  environmental 158, 161, 180
  industrial 98
  occupational 64
  political 155
  quantifiable 137
  surgical 98
  transgenerational 192
  unethical food 119
risk assessment 12, 147, 158–161,
  173, 180, 182, 185, 192, 193,
  195
risk management 64, 91, 125, 157,
  179, 184, 188, 194, 198

ROS *see* reactive oxygen species

safety 2, 156, 158, 159, 161, 173,
  179, 180, 182, 184, 185, 187,
  189–191, 193
scientist 69, 111, 113, 114, 116,
  157, 167, 182, 189, 191
SDGs *see* Sustainable Development
  Goals
separation 32, 44, 209–211, 215
silicate 213–215
soil 65, 68, 74, 75, 78–82
spinning 22, 32, 37–41, 48
stakeholder 118, 119, 121, 126,
  144, 182, 187, 191
standardization 2, 3, 158, 159,
  169, 174, 182, 186, 189, 190,
  192, 194, 196–198
standards 144, 145, 182, 189, 190,
  194, 195, 197
  animal welfare 119
  environmental management
    194
  global 155
  international 193, 194, 197
  labor exposure 189
  mandatory 190
  non-mandatory 192
  quality management 194
  scientific 144, 183
sustainability 132, 134, 135, 138,
  139, 143, 144, 171, 173, 183,
  197
Sustainable Development Goals
  (SDGs) 132, 135, 144
synthesis 4, 6, 10, 65, 79, 151, 155,
  160, 190, 210
system 2, 32, 39–41, 49, 65, 67, 81,
  90, 103, 166, 168–170
  aquatic 79
  artificial ventilation 35
  biological 12, 64, 66, 168, 169
  biotic and abiotic 193
  capitalist 120

cardiovascular 94
deflection coil 9
democratic 125
diagnostic 171
drug diffusion 7
drug release 104
economic 185
education 126
endocrine 66
gastrointestinal 93
global regulatory 178
hormonal 66
immune 68, 155
inner solar 214, 215
legal 135, 185
lymphatic 64, 66, 67
nanofiltration 209
nanometric 172
nanoscale drug delivery 169
patient monitoring 171
physiological 83
quaternary 44
terrestrial 79, 80, 82
therapeutic 168, 169
vehicle-based 208
water purification 75

technique 5, 9, 10, 22, 32, 35, 37,
    39–41, 44–47, 49, 168, 170,
    171
analytical 79
conventional 39
microlithography 210
nanofabrication 211
spectroscopic 10
versatile 10
technological development 119,
    132, 133, 135, 137, 145, 179,
    181, 182, 185, 209
technology 1, 2, 8, 32, 33, 63, 64,
    109, 110, 112, 114–120, 122,
    126, 128, 131–134, 155–159,
    167, 168, 170, 180, 181, 211,
    212

agrifood 119
cold food preservation 117
colloid 113
drug-delivery 168, 169
gene 117
industrial 186
production 181
recombinant DNA 178
template synthesis 22, 32, 41, 42,
    48
textiles 37, 43, 136, 178
therapy 169, 170, 172
threat 144, 161, 183, 186
tissue 7, 12, 31, 32, 38, 45, 64, 66,
    67, 96, 103, 159, 166
biological 37
complex 207
human 92
living 152
lung 90
neural 207
TLC *see* total lung capacity
total lung capacity (TLC) 96–100
toxicity 5, 7, 61–64, 66, 68, 74, 76,
    78, 80, 82–84, 152, 154, 156,
    167, 169

uncertainty 66, 111, 113, 123,
    137, 139, 156
scientific 66, 183, 194, 198
unquantifiable 137

values 98, 100, 102, 110, 115, 138,
    140, 142, 170, 171, 185, 211,
    212, 214
van der Waals force 16, 80
virus 3, 8, 14, 35, 63

waste 65, 78, 213
nanotechnological 146
water 12, 13, 45, 46, 65, 67, 68, 78,
    79, 81, 153, 168, 212
clean 156
distilled 44

welfare 170, 183
   animal 127

world leader 189, 190
world market 160, 177, 189, 194